PHANTOM WORDS

A STROKE VICTIM'S LOSS LEADS TO AN AMAZING GAIN

TED ODOM WITH BRENT ZWERNEMAN

Bloomington, IN   authorHOUSE™   Milton Keynes, UK

AuthorHouse™
1663 Liberty Drive, Suite 200
Bloomington, IN 47403
www.authorhouse.com
Phone: 1-800-839-8640

AuthorHouse™ UK Ltd.
500 Avebury Boulevard
Central Milton Keynes, MK9 2BE
www.authorhouse.co.uk
Phone: 08001974150

First published by AuthorHouse 7/14/2006

ISBN: 1-4259-3933-3 (sc)
ISBN: 1-4259-3932-5 (dj)

Library of Congress Control Number: 2006904645

Printed in the United States of America
Bloomington, Indiana

This book is printed on acid-free paper.

DEDICATION

To my parents, Ed and Toy Odom, for their unfailing love and help in the stroke's aftermath. To Dr. Jean Foster, my sparkling speech therapist who taught me English for a second time. To Dr. C.R. Creger, whose overwhelming patience enabled me to work once more at Texas A&M University. To my friends and family, whose prayers and support lifted me through the tough times. Above all, to God and His Son, Jesus Christ, Who make all things possible.

Ted Odom

To my older and mostly wiser brothers, Stanley and Wesley Zwerneman, extraordinary role models and wonderful family men – and one-time great athletes who taught me a love for sports.

Brent Zwerneman

TABLE OF CONTENTS

ACKNOWLEDGMENTS

Thank you to Bryan-College Station Eagle publisher Donnis Baggett for introducing Brent Zwerneman to Ted Odom. Thank you to Eagle managing editor Kelly Brown for lending an eagle eye to the editing of the Phantom Words.

Thank you to Ed and Toy Odom, Lauri Odom Edles, Pat Odom, Jean Foster, Walter Bottje, Bill Krueger and Howard Smallowitz for providing the invaluable "Other Voices" portion of the Phantom Words.

Thank you to the managers at the San Antonio Express-News for their understanding of this project.

INTRODUCTION

On my first venture to Dr. Ted Odom's front door on Todd Trail in College Station, Texas, in the spring of 1997, a turtle huddled in Ted's shrubbery a few feet from the doorbell. The little fellow quickly clamped up as I stood at the door and somewhat nervously anticipated meeting this professor – a man later best known simply as Ted – with a remarkable story. I might have hopped in with that turtle had he room in his shell.

As a sportswriter, I hadn't always read the newspaper outside of the sports page. But an article in The Bryan-College Station Eagle on Christmas Day 1994 had caught my undivided attention, starting with a large, glorious picture of a fellow named Ted Odom, handsome enough to play a TV doctor.

The photographer had cast a green electronic scan of Ted's brain over his shoulder in the picture, creating a fascinating image of a man who'd beaten great odds in the year following a serious stroke. Now, standing at his front door the fear of how we'd interact – really, just get along – quickly melted when a bright-eyed fellow with a firm handshake greeted me and invited me inside to talk about the possibility of writing a book on his recovery.

Ted, a poultry science professor at Texas A&M University, had progressed from not even knowing his name following the stroke at age 41 to once again teaching on the campus he so dearly loves. As we chatted Ted sometimes paused, searching for a word, and such a consequence – especially in the excruciating few months following the stroke – is why he titled this book as such.

"I had echoes of things," Ted says of his lost ability to communicate. "And they were echoes of things I couldn't scratch."

William Haber of the New York University Bellevue Medical Center writes, "Practically every person who loses a limb or part of an extremity experiences the same curious sensation – that the lost limb is still there." Ted experienced the same frustrating sensation with a once-familiar language that in the span of a few life-threatening minutes jettisoned beyond his grasp, hence the "Phantom Words."

Ted has continued his extraordinary recovery in the years following, and with a positive and bright outlook those closest to him say he rarely possessed before. Here is his story, including parts of the chapters called "Other Voices," where family and friends offer insight into Ted's struggles and triumphs against the nation's leading cause of long-term disability.

It seemed that frightened little turtle and me – both so afraid of the unknown – had much to learn from this extraordinary and kind man.

Best Wishes, Brent Zwerneman

PREFACE
By LAURI ODOM EDLES, Sister

Ted is my hero. His determination in overcoming his stroke is inspiring, as Ted relearned the language and grew into a new person in the process.

Since his stroke my brother's generosity has multiplied, he volunteers much of his time and his spirituality has grown immensely. He's learned to enjoy and appreciate life more intimately.

A couple of months after Ted's stroke in March 1993 my parents asked me to come in from out of state and help him straighten out his financial and insurance issues, as I'm a lawyer dealing in such things. The person who greeted me at the College Station airport – his girlfriend had driven him there – looked like the same Ted, he just couldn't find the words to express himself.

But I also could see a change in the way he cared about me. Every day on my visit he prepared me lunch, something he wouldn't have dreamed of before. One afternoon he wanted to know if I liked tuna, but he couldn't remember its name.

"Do you want," he asked before pausing, "tongue for lunch?"

Ted's kitchen filled with laughter when we both figured out what he was asking. Humor has played a big role in his recovery, which has helped his spirits and the spirits of those around him, along with the depth of his recovery. He can laugh at his sometimes inability to find a word, or at mistakes in his grammar.

One evening my husband, Joe, was chatting about college football on the phone with Ted, and Ted couldn't remember the name of the university in Norman, Okla.

Suddenly Ted burst into the title theme of the musical, "Oklahoma," and that association brought back the name. Ted and Joe laughed heartily at Ted's imaginative means of finding the right word.

Ted is a wonderful example of how one can create a new and better life from a major setback, and in fact find many hidden blessings.

Ted Odom.

CHAPTER 1
Signs of the Storm

I didn't know I was a professor. I didn't even know my name. But I knew I was dying while standing on that bright white tile floor, sharply dressed in a dark blue suit and blood-red tie, only seconds before busily explaining to fellow scholars why chickens have hypertension of the lungs.

I don't know how long I was out of my body in the New Orleans Convention Center on that critical day, March 29, 1993. Was it three seconds, a minute or five minutes?

I just couldn't believe I was dying. I was a 41-year-old, energetic scientist at Texas A&M University with many things left undone. Only moments before I'd been quite certain I'd dwell in this world the better part of forever, despite a rash of complications that foreshadowed this severe stroke.

Now, as I gazed from above at my soulless physical being, I knew with all of my heart an abrupt end to the better part of forever was breathtakingly near. Suddenly – and oddly considering the wild unknown of such an experience – I craved the peace just beyond the grasp of this near-death flash.

***

The problems leading to my eye- and soul-opening stroke, one prompting a prolonged personal struggle to someday teach again, began

1

seven years earlier, in the spring of 1986. I lay sprawled in my atrium in my comfortable home in College Station, Texas. A nice, sunny Sunday had greeted me and I was tanning, resting, listening to some tunes on a radio station touting itself as a better mix of music.

"Good afternoon, y'all," the disc jockey practically hollered over the airwaves. "It's another beautiful day in the Brazos Valley."

But when I opened my eyes to enjoy these cozy surroundings, double vision and dizziness slapped me. I hurried into the bathroom and stared in the mirror. I gazed at my blurry face as my right eye shifted out and stayed out. Vertigo – the startling sensation of dizziness – engulfed me and I couldn't step straight.

"If I lay down," I calmly thought, "whatever this is will pass."

But stretched on that couch and with fright creeping into my heart, I couldn't relax my eye. A few minutes later I was stretched out again, this time in the rear of my 1986 yellow Toyota van, the area usually reserved for two Rottweilers, Kato and Morgan.

My girlfriend at the time, Linda, and I sometimes displayed the pups at a couple of local dog shows, and that's where they rode in an infrequent search of canine fame and glory.

It wasn't the first time I had cruised in the back of the van, sprawled on my back while the Toyota sped down the hot asphalt of South Central Texas. Once, I drank so many margaritas while partying on Sixth Street in Austin that Linda drove the two hours back on the country roads to College Station.

Now the dogs' haven once again became mine as Linda gripped the steering wheel and navigated the streets of College Station to nearby Bryan, Texas, and St. Joseph Hospital.

Linda, a nutritionist, once had told me, "You have an eating disorder," about my strict diet of "lean everything" geared toward running marathons. I lived on chicken, salad, pasta and baked potatoes and little else, because I counted calories like a no-nonsense accountant tallies pennies. Now, Linda reasoned that this neglect for a balanced diet caused the wandering eye and vertigo.

In the emergency room nurses ran me through the typical tests – things like bloodwork and checking my glucose and potassium and calcium counts – and by this time my eye had shifted to its rightful place. The emergency room doctor said she figured I was OK after surveying the test results. At 6 feet and 165 pounds, I was lean as a rod and training for the Houston Marathon in January of the following year.

"Maybe," the doctor reasoned, "you're overdoing it."

She might have been right, since my biological speedometer usually clocked 100 miles per hour when I set my mind to achieving another goal. But this double-vision and roaming eyeball thing seemed a bit strange – despite the doctor's assurances – so I made an appointment to visit a neurologist.

The brain doctor, a fellow big on testing reflexes, asked me to close and open my eyes during the visit to his office in Bryan. Fair enough, but then he asked me to extend my hands and try and hold my fingers up as he pushed them down. Try this at home. It's a tough thing to do, and my fingers naturally dipped.

The neurologist jumped to a diagnosis from the dipping digits.

"Well," he sternly said, "you have myasthinia gravis."

Myasthinia gravis? This was a disease characterized by muscular weakness that mostly affected the face, tongue and neck, and one that usually led to death.

"I don't think so," I quickly countered, stunned by the diagnosis. Me?

"It's only going to get worse," he said.

Then he leaned over his desk, his coat nudging the maple finish.

"Do you want the good news or bad news first?" he asked.

I had liked to think I was a tough guy, so I swallowed hard and asked for the tough stuff first. Slowly and deliberately, he said the worst thing is you'll die from the disease – but he didn't offer a timetable.

"But you'll be on a respiratory machine," he said with some degree of comfort, "and you won't have to take a bunch of drugs."

The latter, of course, was the good news. Talk about your big hits in life. I could run and swim for hours and bike for days, and now I'd been pegged with a muscle disease. Grimacing, I pictured my body disintegrating to the point of needing a wheelchair – and then worse. The neurologist roamed dismally for sympathy.

"You can still do all of those things for now," he said, "but I'm going to follow the evolution of your disease."

I envisioned life minus vital functions – one of helplessness and depending on others when I'd never needed to before. I was a control freak, and I ruefully was learning you can't control all of life's rugged twists.

The doctor ordered more tests to be certain of his diagnosis, and these required multiple doses of drugs hardly healthy for your system. Back in the home we shared on Todd Trail, Linda and I mulled taking the additional tests.

"I don't think I'm going to have it done. The last thing I want to be is dead on some doctor's table," I said, graphically picturing something going awry with the multiple-drug injection. But I finally settled on undergoing the procedure when Linda agreed to go with me.

Once I was at St. Joseph, nurses connected a tube to the middle vein in my left hand. That process, which injected a bright green dye, lasted about 30 minutes and it hurt. Bad.

Certainly, later there were other procedures more painful, but that vein really inflamed after the test. I could feel it protruding in my arm and I could see the stark vessel because it practically glowed a ghastly green.

I rubbed an inflammation from the test for about six months, about the only result of the procedure. The doctor didn't determine anything further from the test, and he didn't change his diagnosis. By his account, myasthinia gravis gripped my body, and in my late 30s, time clearly wasn't on my side.

CHAPTER 2
Life As I Knew It

My first portrait of Texas A&M and its surroundings – painted by a Texan and my major adviser at the University of Illinois in Champaign-Urbana, Paul Harrison – wasn't particularly flattering.

Dr. Harrison was from La Grange, Texas, a picturesque town on the banks of the Colorado River in the gentle hills between Austin and Houston. And College Station, well, let's just say its most defining physical characteristic may be Kyle Field, a nine-story concrete monstrosity and home of the Fightin' Texas Aggie football team.

But as a graduate student at Illinois, I had an interview in College Station set for August 1981 – from information on a flier on a hallway billboard at Illinois – and planned to fly into "Aggieland" to check into an assistant professor position in the Poultry Science Department.

So I asked Dr. Harrison about College Station, which is an hour northwest of Houston, as we hiked the halls of Mumford Building, home of Illinois' Animal Science Department.

"There's not a lot of hills or anything," he said before adding for good measure, "and it's not very green."

I flew to Dallas from St. Louis and then picked up a two-prop "puddle-jumper" from Dallas-Fort Worth International Airport to Easterwood Airport on the far west side of the campus.

"It's not that dry," I thought as I gazed down at Bryan-College Station through the window. "It's actually pretty green."

But Dr. Harrison was right on at least one count: there were few hills. As I climbed in the car of Bill Krueger – the then-head of A&M's

Poultry Science Department who picked me up at the airport – he laughed as I quickly rolled down the window.

"Raise 'er up," he boomed in his best Southern drawl. "Everybody has air conditioning in Texas, even in their car."

So went my first lesson in Texas etiquette. After an agreeable interview in College Station, I flew to the University of California-Davis, which is near Sacramento. Talk about your changes in latitudes and attitudes. There, officials just shot me through a few interviews and pitched Davis to a near nauseating degree.

"We have big names here, you know," one interviewer told me.

Even then I was thinking about Texas A&M and how its people prided themselves on friendliness. Everyone in College Station crooned "Howdy" this and "Howdy" that, even when you strolled into an off-campus eatery like "The Chicken Oil Company" for a half-pound cheeseburger with grease deliciously dripping from its edges.

"Howdy" is the signature of Texas A&M, and that greeting starkly contrasted the seeming stuffiness and isolation of Cal-Davis. Upon my return to Champaign-Urbana from Davis a letter from Dr. Krueger waited in the mailbox, offering me the A&M job.

I accepted in a heartbeat. I was 29 and recently divorced from a marriage of three years. In my late 20s I had focused on trying to finish my Ph.D., and my then-wife, Carol, who's an artist, had wanted to move to Chicago so her paintings might receive more exposure. We tried justifying the divorce by claiming at 26 we were too dumb and too young to have married, but really that was a lousy excuse and a cop-out. My parents were married when they were 18, and they learned to grow up together.

On the September night our divorce became final Carol and I "celebrated" with champagne in a bar in Champaign.

"Our divorce," we gleefully said when fellow patrons asked the reason for our hearty toast. Some laughed, others didn't. Many people just shook their heads and considered it an odd festivity. They were right.

Shame coursed through my veins that marriage hadn't worked out for Carol and me, and I didn't even tell my parents we were divorced until after Christmas, more than three months after it became final.

During a Christmas banquet in 1980 for all of the kids at the Epworth Children's Home my father runs in St. Louis, I was perched at the head table enjoying the festivities, when my dad introduced me.

"This is my son," Ed Odom's voice thundered from the podium. "His wife, Carol, couldn't attend this evening."

I thought, "Oh, boy, I've got to talk to my parents."

A couple of days later at a Christmas party for the home's workers, a woman sitting next to me asked, "So, how is your wife?"

Sipping on a Budweiser beer in a bowling alley bar, my shoulders sagged a bit as I explained my predicament.

"I haven't told my parents yet," I muttered, "but I'm divorced."

****

I met Linda at A&M's Kleberg Center – home of the Animal Science Department – a few months after beginning my job as an assistant professor in January 1982.

She was separated, and a year after we met her divorce became final. We'd often cruise to Austin, spending nights on boisterous Sixth Street, drinking and dreaming.

"We should get married and have children," I'd say playfully, and we'd laugh over margaritas or whatever drink or mix of drinks we preferred that night.

But flimsy and false hopes formed that life's foundation, and most of such banter was only the alcohol talking. Usually on the drive back to College Station depression engulfed my entire being, because inside I knew God hadn't planned this life for me.

For three years I lived in an apartment complex in Bryan, before I bought a three-bedroom house in 1985 on Todd Trail that's still my home. About that same time Linda moved in, and though I sporadically attended a Methodist church in town, that didn't keep us from living together.

I was teaching A&M undergraduate students in Environmental Physiology and graduate students in Avian Physiology, and in the meantime ignoring my own low self-esteem. Many people try to be perfect – although that's impossible – and I was one of those individuals. An uneasy tension bubbled just beneath my skin, sometimes surfacing like molten lava.

"Did you even read the book?" I'd yell at students when they sat in class with quizzical expressions on such matters as a bird's heart or kidney structure. "If you studied and read the book, you wouldn't have so many questions."

Frustrations like these racked me for four years, because students just had questions and more questions. It didn't help I was suffering from unbeknownst mini-strokes. One time a student, as students are prone to do, tried extending a test day.

"Can we have the exam on Monday instead of Friday?" he asked. "That would give us more time to study."

I cast an arched eyebrow toward the unsuspecting collegian.

"You're taking the exam on Friday," I snapped. "You shouldn't need the extra days if you were prepared. You're an adult and you should have studied. And I'm not going to waste my time answering every goofy question you have, without your having read the book."

Of course, other students picked up his torch and from then on in that class they goaded me with questions of: "Can we extend the test day?"

Honestly, I didn't know what teaching right meant. You must be extremely patient, and even if students ask the same question three different ways, you must answer it the same friendly and helpful way. The then-head of the Poultry Science Department, Dr. C.R. Creger, pulled me into his office and shut the door one day in the spring of 1986.

"Students don't like your teaching," he calmly said. "You don't teach in ways they understand. You talk on a higher plane and you're not connecting with your students. You're only teaching for yourself."

I deserved such criticism, because the students simply didn't like me. My teaching manner was too intense and I only wanted to lecture, and not take time to answer questions I believed that they should already know the answers to.

I was an angry man and taking it out on my students – and Linda. One day in the fall of 1989 she threatened to move out, and I responded with a defiant, "Yeah, right," followed by laughter. So, she did, and we broke up.

At that point my attempt at escape from reality was long cycles through the countryside surrounding Bryan-College Station. I'd pedal for five hours at a time, down to the Wellborn Community and up through the little town of Carlos and by a huge power plant that lit up beautifully in the evening, all on peaceful backroads that formed an antithesis to my volatile personality.

The entire ordeal I was acutely aware Jesus Christ died on the cross at age 33, and at the same age I truly believed I needed a calling, although I refused to act on it. I just occasionally talked about it, which without action is good for nothing.

"You just want to be a monk at the top of a mountain, thinking deep thoughts all day without a care in the world, don't you?" Linda asked one day not long before she left for good.

Such a setting might have made a nice alternative for this then-lost soul.

CHAPTER 3
The Problems Persist

The neurologist first concluded I had myasthinia gravis in the spring of 1986, and when I visited him again a couple of months later, I jogged six miles on a treadmill so he could test my muscle strength. His deduction: I still had the disease but wasn't showing any signs of weakening.

Remember the finger test?

"That's a crazy diagnosis," one friend said, echoing the thoughts of many others close to me. "No one can hold his fingers up in that test."

But the neurologist insisted my failure to do so meant myasthinia gravis. I didn't bother with a second opinion, because of the hassle involved and maybe because I might not like what the doctor might have said. Instead I just ignored the neurologist's opinion – and also the casual opinion of a neurologist who attended church with my mother and father in St. Louis.

"Ted should undergo lots of tests before such a diagnosis can be made," the brain surgeon told my parents.

I just figured whatever happened, I wasn't going to cry or whine about it. I continued running and doing all of the athletic activities I'd done before, with no sign of the myasthinia gravis.

In the span of the several years I thought I had that disease, I resolved to do everything I planned before my muscles became too weak.

"I'm not going to stare at the wall and be a baby," I told friends.

I tried possessing a promising outlook, but secretly I always wondered what day I would finally feel the physical tugs of the disease. It wasn't a pleasant way to live.

I should have trusted God more. The neurologist claimed I had that disease but in my heart I believed I didn't. I was stubborn to understand myasthinia gravis, and others suggested denial gripped me. Truth is I'm a very trusting person, and one who always followed doctors' orders. I've learned better.

***

In June 1989, three years after I'd first shown signs of trouble, I suffered a sharp pain in my left arm, while in bed in the middle of the night. Waves of anguish ripped through my fingers on up to my shoulder. In the days after I also experienced multiple types of vertigo and dizziness, so I visited an eye doctor.

"I'm not sure what it is," he told me after checking my vision, "but you don't have an eye problem."

I learned to temporarily "cure" the dizziness by leaning over so the blood would run to my head and iron out my vision. It worked if I held it long enough during a dizzy spell, but that surely offered a strange sight to passers-by who had no clue my world was spinning.

One afternoon that year, as I strolled a sidewalk on campus headed for A&M's Memorial Student Center for lunch, I suddenly couldn't control my legs, creating what amounted to a crazy duck walk before hundreds of students between classes.

I quickly turned and zipped back to the sanctuary of my office, where I didn't say anything about the incident. I was too scared and embarrassed.

The continued double vision created a sick-to-my-stomach feeling, but I would shallowly "reassure" myself, "It's just part of the muscle disease." During those episodes I would just bend over and that seemed to work – for a while.

I visited another eye doctor and he prescribed reading glasses. He figured they would straighten out some of my double vision and occasionally they did. But the glasses still weren't a cure, and so now I was bending over while holding my spectacles on my nose. Despite its seriousness, this offered a curious and sometimes comical sight.

The list of doctors continued, while I searched for answers to the ailments. I next visited a cardiologist, who placed me on a treadmill for an echocardiogram. Such a test measures the small amount of electric energy the heart produces.

"You mind if we crank it up?" I asked with a grin. I thought the test was a bit too easy on my system since I always run, but he didn't appreciate my attempt at humor.

That was a Friday examination, and it pained me to wait until Monday to hear the results of a test, especially when it didn't seem like we were making much progress with diagnosis after diagnosis. The cardiologist's matter-of-fact result: My left ventricle was a little thick, from working out and exercising through the years.

"A lot of athletes have the same problem," he said. "That's why they call it an 'athlete's heart.'"

He then told me of a Russian weightlifter who also had a thickening of the ventricle but had continued in his profession with no complications.

"Go ahead and do what you want," the specialist said as my heart suddenly seemed light as cotton.

He called it an "adaptation," and I was familiar with the word because I always saw it in chicken hearts. All I had, it seemed, was a thickening of the left ventricle, a normal thing.

"I'm just an athletic guy," I thought while grinning and practically skipping out of the cardiologist's office doors.

Clean bill of health from the cardiologist notwithstanding, the troubles continued and now included a mild form of aphasia – the inability to remember words. At the time I dismissed the uneasy feeling of forgetting some words as anxiety attacks resulting from my suddenly poor health.

But the idea of groping for words that used to be as ordinary as the air in my lungs haunted me, so I finally returned to the cardiologist with my fears. Following an echocardiogram he only shook his head when returning with the results.

"You have the same thing I told you last time," he said. "And the left ventricle has got the same thickness as before."

He professed he didn't have any idea what might be causing the double vision and vertigo, along with the inability to remember some words. Finally, I visited an internal medicine specialist in Bryan-College Station.

"Next time you have any problems, drive straight to my office and I'll check you out," he said.

That seemed an odd appraisal, since the last thing anyone wants is a dizzy man behind the wheel. The specialist jotted down a few factors of stress and how to cope with such, but I insisted I worked well under stress daily.

In all at least three doctors misdiagnosed my ailments. It seems at least one should have figured out I had suffered some mild strokes before the big one.

I doubt I'd won had I sued any of the doctors following the severe stroke. In court I'd have sat in the witness chair answering questions before the jury, not wholly damaged and therefore not worthy of a settlement.

But only the extraordinary grace of God saved me from a raging stroke that hid around the bend, one at least three doctors never saw coming.

Young Ted practices his football stance

CHAPTER 4
Coming of Age in Illinois

In January 1960 the Odom family moved from Southern California to Illinois when I was 9 years old. Putting it mildly, we didn't own the same heavy clothes as the other kids, and Sears & Roebuck filled quite a large order of snowshoes, thermals, scarves and jackets that first week in our new wintry environs.

My father, Ed Odom, had run a children's home in LaVerne, Calif., and sports had been a natural part of life for us because of all the boys and girls around. Plus, my dad stressed the team concept you learn from games.

That love of sports continued in our new home of Urbana, Ill., and if an Odom was around, you could be sure he or she would be swimming, diving, playing golf, football or about any sport imaginable to pass leisure time. Or at least when the weather warmed enough to permit such activity.

I was the second oldest of five children, and Ed and Toy Odom had three girls and two boys. I'd seek my sisters' advice on the ways of women, because I figured that gave me an advantage over my friends in dealing with the fairer sex. Sisters can be awfully smart, even if you don't fully appreciate it at the time.

At Urbana High School, I took to biology and chemistry and those "science" kinds of classes, but I didn't have good grades. I didn't work hard when it came to school, I mostly did the minimum and had a natural grasp of science because of my curiosity toward such. It wasn't a priority to make an A when I could just as easily graduate with a C.

I was bad about studying for tests and finishing daily homework, and even needed a tutor to keep me from failing Spanish. If something interested me, however, I'd buckle down and complete a sizable task in a day's time. Take piecing together a toy car model, for instance. I was so impatient but at the same time so focused, that I'd glue the parts together before the paint dried.

One day when I was a junior in high school, dad sternly pulled me aside in the dining room. He obviously wasn't going to ask me to play a little catch.

"Son," he said in a tone only a father can summon. "When it comes to school, you've got to work harder. And don't boast about unfinished tasks (I was particularly bad about that, with responsibilities that didn't interest me). You must start thinking about the future."

There's no doubt sports helped me with my future, even if that eventually meant researching chicken hypertension.

During a heated football contest against rival Springfield on a cool September evening in 1968, my father peered from the stands at my play as a tight end for Urbana High. I was particularly proud of my blocking, before dad quickly squashed any smugness on our stroll to the car after the game.

"It's good that you blocked a guy on each play," he said. "But most of the time the plays weren't done when you blocked him. If you want to play in college, you're going to have to get off the grass and go block another player."

The message: Don't let up until the last whistle has blown. When things aren't going exactly as planned in life or it's a particularly rough day in my continued recovery, that one comes in handy because it's all about getting off the ground – and tackling the next obstacle.

At Urbana High, football's Warren Smith ("Coach Smitty") used to claim, "It's better to give than to receive." In terms of a good hit, anyway. I was smaller than most in football and so I mustered every ounce of energy and focus on blocking, and once I laid the blow I quickly focused on the next one.

That can apply to daily tasks. I've learned not to be overwhelmed by all of the blocks I have to execute against sometimes seemingly overwhelming obstacles. When I take it a lick at a time I'm usually amazed by the results.

Ted receives an encouraging hug from Coach Warren Smith during an Urbana
High football game.

Football, for all the lessons it teaches, wasn't even my primary
sport at Urbana. I also learned to dive at my dad's children's home in
California on an old, stiff board with burlap tacked on the end to keep
the kids from slipping. The old plank barely sprang, but it was three
feet off the water and enabled us to sharpen our diving skills.

I even continued diving my first year of college at Eastern Illinois
after earning the district title at Urbana. Diving requires a completely
different kind of focus than blocking in football.

One mental slip in diving and you're flat on your back or belly, and
water seems nearly as unforgiving as dirt from a few meters. So nothing

makes you more focused than diving – for obvious reasons – and the more I practiced the better my balance and concentration became.

When I was 10 during a diving competition, I walked to the end of the board and turned backward to prepare for the leap. But I slipped – maybe I wouldn't have had burlap been tacked on the end – and the judge rewarded me with a nice round zero for my efforts. I cried as I climbed from the pool.

"I'll bet you don't ever slip off again," my dad said with a half-smile.

Seven years later, when I won the district diving championship, one of the judges clutched my hand.

"Ted, I remember when you slipped off the diving board when you were 10," he said with a broad smile. "I guess you learned your lesson."

In October of 2005, and quite surprisingly, I was elected to the Urbana High Sports Hall of Fame. I was thrilled to be honored.

OTHER VOICES
Father, Ed Odom

Ted began diving when he was 8 years old, and in high school he excelled as a pole-vaulter. Because both sports take a tremendous amount of bravery, we weren't surprised at his courage in dealing with the stroke.

Ted was an extremely curious kid who wanted to know how things worked, and he asked a lot of questions. He had some academic problems, which we now believe was undiagnosed dyslexia. Early on I wouldn't have pictured him as an academian because of his trouble with comprehension.

That's probably why we pushed so hard in athletics, because we figured that was his strength. He was a highly intelligent kid, but we never figured he'd become a professor.

Looking back at all of those sports Ted played, you can see that many things were training or preparation for tougher days ahead.

Mother, Toy Odom

Ted was always extremely thoughtful of others as a child. In the second grade he bought a soap-on-a-rope as a Christmas gift for a little boy in his class that was always dirty. Another little boy never had lunch money, so Ted would give him his lunch money, or buy him peanuts.

Ted mostly tried emulating adults. His dad would leave change on the bathroom counter at night, and Ted would pick up the change and take it to school, because he always wanted to have money in his pocket, just like dad.

CHAPTER 5
Happily Blindsided by Science

Following my dad's "don't say it's done until it's done" speech my junior year of high school, I made decent enough grades to enroll at Eastern Illinois University in Charleston – about 40 miles south of home – in the summer of 1969, to gain entrance in the fall.

That fall I declared myself a physical education major, and Cat Anatomy was required for all such majors. Often the best way a person can find his or her niche is to try different things, and a big part of college is weeding out what you don't want to do in life. Finding my niche came in the form of a stiff, cold, dead cat.

With each portion of the cat's dissection, I grew more comfortable and obsessed with the process. Professor Ed Moss noticed.

"Can you help some of the other students?" he asked when I had finished the procedure quickly and neatly. That was my first taste of teaching, and to help others with something that came so naturally sparked an interest that not even a serious stroke stopped.

One day in class Dr. Moss studied me in earnest while I happily explained to another student the different muscle layers in the dissected left leg of a cat.

"I see you're a P.E. major," he said after class. "Why?" I explained it seemed the most natural – I didn't say easy – major for me.

"I like coaching," I said simply. He shook his head and offered, "What about zoology?" I shrugged.

Later, I visited my college adviser, who was an assistant football coach.

"Why did you sign up for these classes?" he asked incredulously of a spring schedule that included Comparative Anatomy and Botany. "I'm not sure you can do this."

I boldly announced that my life was my own, and that I was switching from P.E. to zoology.

"Well, good luck," he said. "But you'll be back."

As I worked on my undergraduate degree I also participated in diving and pole-vaulting for the university. But I wasn't on an athletic scholarship and I didn't have time for the three "S's" – sports, studying and Sigma Pi fraternity duties. So I retired from diving and pole-vaulting to concentrate on the books – and the socializing.

The partying, dancing and dating from those first couple of years all seemed to string together, and while I lived for the moment there wasn't an end goal to all of that fun and shallow self-fulfillment.

One spring day in my second year at Eastern Illinois as I strolled the four blocks from campus to the fraternity house with grades in hand, I figured things had to change. I was about to flunk out, and had lost sight of goals after graduation.

So I buckled down and began studying and visiting the library, a building that had just been a rumor to me early in college. About three years after the visit to the adviser's office ("You'll be back," he'd said) I nearly possessed a zoology degree with a minor in botany.

But a couple of semesters shy of graduation I impishly shucked all of that college for a construction job, something I simply loved – whether it was painting, digging ditches or framing a house. When I painted, especially in the highest point of a roof's pitch, I compared its elegance to gymnastics.

I suspect the Occupational and Safety Health Administration might have had a field day with some of the routines I performed on the top of that ladder, while trying to reach every cranny with the same even brush stroke.

I enjoyed the short, tangible goals that construction offered. If you finished a cabinet, for instance, you could stand three feet from it and admire the handiwork for a bit. There it was, ready for use. A college degree seemed like a distant, intangible thing, and that idea can be tough to grasp for an impatient man, one who as a boy glued his models together before the paint had dried.

Deep down, though, I knew construction wasn't my true calling because of my love for science, and so I returned to school within a year and earned my diploma. A year after graduation I enrolled at Eastern Illinois in pursuit of a master of science in zoology.

By 1978 I had a master's degree and became intent on obtaining a doctorate. My old Boy Scout leader, Larry Larimore, was a wildlife professor at the University of Illinois. He relayed to Paul Harrison there was a young man he knew who was interested in animal science, and Harrison – a professor in the department – granted me a partial scholarship to enroll in Illinois' program.

Looking back, I'm sure I was No. 1 on everyone's geek list that first few weeks of chasing down my doctorate. While others wore jeans and T-shirts in pursuit of higher education, I haughtily sported slacks, button-down shirts and even a briefcase instead of a backpack. Remember, this was the freewheeling 1970s, and I may as well have had "NERD" stamped on my forehead.

My image of higher education was as snooty as my appearance, and since I was shooting for a Ph.D. I believed I needed a wardrobe suitable for a doctor. My former classmates probably still giggle about that.

When I finally switched to jeans and T-shirts when I realized a necktie and pressed shirt wasn't going to earn a doctorate any quicker, everyone breathed a little easier – especially me.

One of my projects in college was following through on an idea of Dr. Harrison's that chicken shells might become harder if the chickens ingested carbon.

The idea was to feed the chickens carbon dioxide through special feeders in their coops, and see if in fact the shells came out harder. It was an innovative thought and one that the wire services and newspapers picked up on because of its uniqueness and the innate humor of chickens gulping carbonation.

I even chatted with a few radio shows in Boston, Chicago and San Francisco about the project and Omni and Science '82 magazines ran articles on the experiment. So went my self-proclaimed 15 minutes of fame.

OTHER VOICES
Excerpt from Science '82 Magazine
May 1982, By Howard Smallowitz

Every year $300 million to $600 million worth of eggs never make it to American frying pans because they are cracked. Now two animal physiologists have found that carbonated water can significantly increase the strength of the eggs that chickens lay. It may not be Perrier, but the trendy hen of the future may be quenching her thirst on sparkling water.

In hot weather chickens pant to cool themselves. In the process their bodies lose a great deal of carbon dioxide. Though most animals exhale carbon dioxide as a waste product, it is important for lots of body processes.

For that reason chickens – as well as humans and other animals – need to keep a constant supply of the gas on hand, according to Ted Odom, an avian physiologist at the University of Illinois. "If this level is changed," says Odom, "many things are affected."

For instance, the amount of carbon dioxide in a chicken's blood helps determine the pH – that is, how acidic or alkaline the blood is. When a chicken pants heavily, the level of carbon dioxide in the blood falls, and the blood becomes more alkaline. "The blood's pH has to be maintained within a pretty narrow range," says Paul Harrison, a professor of science at the University of Illinois who collaborates with Odom. "To compensate for the greater alkalinity, the kidneys begin to excrete bases," or alkaline substances. Several of these bases are compounds containing calcium, a major component of eggshells.

Another of the main ingredients in eggshells is carbonate, which hens make from carbon dioxide. Thus as carbon dioxide levels fall, the egg layers have less raw materials in their bodies to work with, and they begin turning out fragile eggs.

Efforts to put calcium in the chickens' feed proved unsuccessful. With the blood already high in bases, the kidneys quickly excreted the additional calcium. Attempts to increase the concentration of carbon dioxide in the poultry's air proved difficult because it required making the chicken coops airtight; high levels of carbon dioxide also gave workers headaches.

Then Harrison and Odom slipped carbon dioxide into the chickens' water. The alkalinity of the chickens' blood went back to normal, thanks to the sparkling water, and their kidneys were no longer compelled to unload calcium. There was also plenty of carbonate available to devote to shell making.

The resulting eggs are about one-half of one percent heavier than eggs laid by hens who drink uncarbonated water. "This may not seem like very much," says Odom. "But there's a very fine line determining whether a shell will break or not. We believe that we've passed over that line."

CHAPTER 6
Struck in the Big Easy

By 1993 I had a girlfriend of three years, Brooke, and the same ol' medical problems – vertigo, dizziness and double-vision – that climaxed in the March stroke of that year. A year earlier Brooke, who worked as a medical technician in a hospital, had mentioned her theory on my ailments.

"I think you've suffered a CVA," she said.

That's a cerebral vascular attack and characterized by blocked vessels in the brain. I returned to the cardiologist and asked about the possibility of the problems stemming from a CVA.

"You're too young to have a CVA," he said. "Those are for people who are much older. You're just stressed out."

I then told my psychologist, whom I was visiting because of relationship problems with Brooke, that I believed I had suffered a mild stroke.

"No," he said, "anxiety is the reason you can't talk."

Later that November 1992 week I was scheduled to present in Delmar, Maryland, on genetics, nutrition and pulmonary hypertension in chickens at a nutrition symposium, and I asked my psychologist what I should do if the words don't flow.

"Only you know if you're having an anxiety moment," he said. "If you come to a slide and you don't know what it's about, skip over it. No one will know."

So it was click, click, click at the Nutrition Conference in Delmar, as I hopped, skipped and jumped over slides that had slid from my memory. Despite the bullet-riddled report my colleagues told me, "Good job," as I learned to quickly cover my growing forgetfulness.

It wasn't any better in the classroom or in my office, where Texas A&M students visited and sometimes left shaken by the experience. Once a rather meek student wanted to discuss physiology, and so I invited him into my office. There, I just sat in my chair and stared across the desk, forgetting what to say. It was frightening – certainly as much for him as me – because no words were exchanged. I was completely frozen, and he expected me to speak first.

I now better understood those classic "Honeymooners" episodes when Ralph made silly gestures and Alice just sat there, not paying him any mind. I felt like Ralph, as that poor kid just sat as I stared at him and he peered at me. That lasted a couple of minutes, or an eternity in this bizarre staring contest.

Finally, he rose from his seat and hurried out with nothing. That type of exchange – or lack of – was a first, and extremely unsettling.

Then there was class, where I began to feel the same strains of a seventh-grader who can't remember the details of his book report before a bunch of critical mates. Once I was presenting on calcium in the blood and just stopped in mid-sentence. Finally, I broke the tension when I remembered I had corrected an exam.

"That's enough for today," I said to the relief of some and arched eyebrows of others. "Let's go over the test."

I had dodged a bullet – this time.

"What am I going to do," I wondered to myself, "the next time there's no exam to look over?"

Later, several of my graduate students said something had seemed wrong in the couple of months leading to the stroke, but they figured it was girlfriend problems or such. I had lost my focus, all right, but it had zero to do with Brooke, and I had never encountered such helpless frustration.

I also had never used notes when lecturing, because I felt as familiar with the subjects as behind the wheel of my trusty old Toyota. An A&M undergraduate class typically held about 35 students, so I'd stroll around, maybe plop down on a desk and jazz up a lecture a bit to try and keep the students from dozing.

Once I lectured for 2 ½ hours without missing a beat. I felt like a good preacher in that most of my students – my congregants, if you will – connected with my message.

Funny, but I didn't know the term "aphasia" until after my stroke. Aphasia means you lose information. The words or sentences simply drift, and it was happening to me. And suddenly, because of the aphasia attacks, I no longer was the preacher who always stayed in rhythm.

Through the Christmas season of 1992 and on in to January and a new semester and a new class, my momentary memory loss worsened. Plus, in late March I was to attend a FASEB – the Federation of American Society of Experimental Biology – conference in New Orleans, where I'd see thousands of peers.

I became stuck and lost in mid-sentence while lecturing more frequently, and I doubted my ability to teach. Life, to put it mildly, had become flustering.

My personality didn't allow for any real self-probe of what might be wrong. I figured the problems one day would simply clear and life would go on. Such focus and single-mindedness can help keep your life in order, of course, but you might miss out on truly living in the meantime – and you might also miss vital warning signs that something is seriously wrong.

Where did I go wrong medically in the years before my stroke? I should have truly focused on God's Word to guide me through the rough waters, and I should have trusted my friends who insisted that I had problems deeper than supposedly diagnosed.

***

I flew from College Station to New Orleans on March 27, 1993, and was quite pleased with a hotel room at the famed de la Poste. The hotel, tucked between Bienville and Conti streets at 316 Chartres Street, is in one of the oldest and most historic parts of town. The New Orleans Convention Center, where the FASEB conference was taking place, is seven blocks from the hotel.

I got into town on a Saturday, and watched some of the NCAA basketball tournament because I was a fan of then-Indiana University coach Bob Knight, whose Hoosiers lost to Kansas that day in St. Louis for the right to play in the 1993 Final Four.

Once the game was over I headed out for dinner and then Bourbon Street and, man, did I have a wild, reckless night. I drank so much beer I thought my stomach was going to explode.

I strolled Bourbon alone, back and forth on the seven-block sin strip. I wandered in most of the bars, but tried avoiding the cross-dressing clubs. Such evasion is hard to come by on Bourbon, though, because you typically get an eyeful of oddity just hanging out on the street. Around 1 a.m. I downed a couple of hot dogs and headed for the hotel room.

On Sunday, I finally rolled out of bed about 10:30 a.m., and later that evening I visited Pat O'Brien's Restaurant with some friends. But I didn't drink because my presentation was on Monday and the residue of a crazy Saturday night still lined my stomach – and my mind.

All things considered – the continued puzzling health problems highlighted by memory loss – that Monday morning seemed like a really good one, as far as Mondays go.

The meeting consists of about 5,000 presentations, so it's no small affair. Multiple major fields – ranging from physiology to neurology – are on display and some of the best researchers in the world attend.

I arrived at the convention center so early that morning they wouldn't let me in the building. I was to present from 11:30 a.m. to 1:30 p.m. and I had my poster set up well in advance. A poster is exactly as it sounds, a large poster board chock full of information on your research.

Lines and lines of posters blanketed the convention center floor, and when I began studying some of the others, the problems started. My anxiety grew as I couldn't comprehend parts of the posters' text, and my heart rate inched higher.

By then it was about 11 a.m., and I reasoned sitting in front of my poster might be the best thing to do, and maybe my mind would clear. That's when my former adviser from Illinois, Paul Harrison, passed by and began studying my data.

"The only way to be sure the circulation in the lung is correct," Dr. Harrison said in a manner meant to improve my work, "is to inject dye so you can measure it. That way you can see if the circulation is completely open or closed, so the results will be more certain."

He was only trying to be helpful, but my then-thin skin hardly allowed anything close to criticism of my work. That simple exchange with Dr. Harrison was my last "normal" conversation before the stroke, because some true craziness lurked ahead.

I ventured to a coke machine – trying to clear my mind – and when I returned to the poster my co-author, Bill Chilian, stood near with a colleague.

"He'd like to hear about the data," Bill said of his friend.

Bill's buddy was European with a punchy accent, and Bill excused himself while I began explaining a technique to measure the pressure within chicken lungs by pushing fluid through the organs.

"There's not as many vessels in chicken lungs, and because of the weight of the lung the pressure's very high," I told the foreigner. "The idea is the chicken lung doesn't grow at the same rate as the body weight. There's a mismatch in the number of vessels of the lung, which translates into high pressure. And heart failure resulting from this high pressure, of course, induces sudden death."

Or at least that's what I thought I was telling him. An echo of my own words pierced my brain, an altogether new and disturbing experience. But I talked automatically because of my familiarity with the text. The foreigner nodded politely. Meanwhile my brain kicked into high gear, nearly spinning out of control.

"What's going on here?" I thought. "I've got to finish this and get the heck out of here."

About that time Bill returned.

"Ted, why don't you grab a bit to eat," he said, eyeing me a bit strangely. "I'll watch the poster until you return."

"No," I snapped back. "Go do what you need to do. I'm fine."

Another of Bill's friends happened by and Bill said he'd see us a little later. This time I could barely understand my explanation from the start, and this guy – an American – looked more confused than before I began. He quickly picked up that I had a serious problem.

At that point clots bulging in my left ventricle raced toward my brain, and the more anxious I became as my heart rate climbed, the more each beat cleared the clots from my arm.

That's when my world darkened and my mind escaped my body. I saw my physical being from a few feet above, standing with Bill's perplexed friend.

"I'm right here," I thought in awe, "but I'm also right over there."

I was staring at the shell of myself, but I couldn't approach it. I had no physical feeling, as you would with a body.

"This is it," I thought, strangely growing peaceful among the chaos that moments earlier had racked my brain. "I no longer have to wonder how I'll die."

I fully knew I was dying and I wasn't shocked. A clear understanding that I teetered between life and death enveloped me. From behind, a warm blanket of serenity overwhelmed my senses, and a powerful feeling to fall back and surrender to the tranquility engulfed me.

"It's OK, it's beautiful," the serenity beckoned. "Just lay down."

But an overwhelming urge – knowing with all of my heart that God needed me on earth for a rebirth and a life of His service – kept me from that final, peaceful surrender. I rushed back to my body, where noise and babbling suddenly drowned the perfect peace of moments earlier.

***

Before the stroke I pushed myself to understand scientific interpretations of the meaning of life, not so much those based on religion or faith. I didn't read The Bible or The Upper Room – an invaluable daily devotional guide – or anything similar.

Really meaningful prayer or meditation wasn't important to me. Instead after I ran for an hour or cycled for a couple of hours, the hasty send-up to Heaven rattled something like this: "Thank you, God, I made it again."

When I was a young boy, just about every Sunday I was in the United Methodist Church in LaVerne – interestingly enough the backdrop for the wedding scene in Dustin Hoffman's "The Graduate" – dressed up and attentive to The Word.

My parents later said I was like a "Little Minister," for my early holiness. Somewhere along a path that widened the longer I lived, I lost that sensitivity toward God and others. Like many scientists, to be certain of something I needed hard data. Evidence. Proof. Something I could measure.

But during the stroke and since I've jumped from belief in hard data to faith and trust in God, which there's no doubt in the end will prove more rock solid than any of earth's hard data. The proof is in The Bible, and I'm not arguing with it.

Older people are closer to God. Why? Because they have more time for Him, and a better ability to concentrate on what He offers: Eternal Life.

The telephone line from God is always open, and the elderly are for the most part less busy and have more time to answer. As for the rest of us, God too often gets a busy signal.

Heaven can be an abstract concept, for sure. Before my stroke I wanted to touch Heaven, feel it for proof of its existence. In a sense, I got what I asked for.

And part of the Heavenly effect during the stroke was the very real sense of Mam-ma, who as a loving grandmother was a wonderful influence on my life. Mam-ma – Clara Opal Odom – had died in January 1980.

When I was a child, I always looked forward to staying at the LaVerne home of Mam-ma and Pap-pa – James Frank Odom – because of the love for their grandchildren.

The aromas while waking up in the morning weren't bad, either. Every morning before a big breakfast of biscuits, sausage, gravy, eggs and bacon and the like, Mam-ma slowly recited the Lord's Prayer before Pap-pa left for his job as a gardener. It was a simple but wonderful and reliable moment, and one engraved in my memory for its uncomplicated splendor.

During the stroke as I floated free of my earthly body, I felt the kind spirit of Mam-ma and, yes, a touch of Heaven. The incredible warmth that surrounded my body was pure love, and something words can't accurately convey.

The best thing I learned from the experience? There's no reason to be scared of death because of the beauty and eternity behind its doors.

I tried succumbing to the feeling of warmth, and I dearly wanted to lay back and rest and let it surround me. But God gave me the knowledge and a clear message to return and go forward, because I have a purpose: to let people know there's no pain in Heaven, there's only pure love and no reason to fear the unknown.

People who walk with Jesus have glows about them. They're caring, passionate and sensitive. I try daily walking with Jesus, so others see this and give glory to Him. This isn't automatic. You must work hard but the rewards are great.

I'm still learning, too, about faith and trust in God and I'm far from unique in my approach to daily living that gives God the glory. I've changed my behavior every step of the way, as God has directed me.

And opportunities to serve Him abound.

Ted's Mam-ma, Clara Opal Odom

CHAPTER 7
Back Again

I was back again, standing on that convention center floor, suddenly free of the clots that had ravaged my brain. I didn't know how much damage had been done. I believe the clots dissolved the moment God sent me back, although quite a struggle in the next few hours, days, months and years loomed ahead.

Confused and disoriented, I wandered for the exits with the only intent of getting back to my hotel room. A neurologist later explained a hurt animal goes to a place it's most comfortable. Mine was the hotel room and I was desperate to get there.

"Ted!" a good friend, Guy Gardo, called after me. "Ted!"

I ignored him and hurried through the doors. I was in shock, but I had a general idea where my hotel room was – seven blocks from the convention center.

There were small streets and wide, busy streets to cross. I could only see out of my left eye – the swelling from the stroke caused my vision to frighteningly narrow – and when I came to the intersection of Decatur and Bienville streets I hurried right onto the asphalt.

It was late morning, so there wasn't heavy traffic, but there was enough to set off a flurry of horns.

"Crazy drunk!" one bus driver yelled.

Had I been struck by a car and died there on Decatur Street, would my family have even known I'd suffered a stroke? I found the hotel and once in the room, collapsed on the bed.

"If I can just lay down, I can shake this," I stubbornly and foolishly thought.

I was backed in a corner, like a scared cat, and trying mightily to claw my way out. I fell into a fitful rest, and slept for about an hour. When I woke up, the old technique of bending over to clear my mind no longer worked. Through it all, I pointlessly remembered needing to pick up my poster by 5 p.m.

So I journeyed across the busy streets again, and it was honk, honk, honk from the cars the entire way. My vision was extremely narrow and I didn't grasp the idea of walking on the sidewalk, so I spent most of the time in the road. I followed visual cues to the Hilton Hotel next to the convention center, and there I plopped into a chair in the lobby.

The episode reminded me of running in marathons, when you hit the proverbial wall and you're supposed to bust through. That was mostly physical, though, and much less scary than the mental part of this trauma.

I waited a couple of hours in that chair, and found comfort in it because it had grown familiar and served as security. About 5 p.m. – I grasped the time by the hands on the clock – I headed for the poster and as I grabbed it a good-looking woman from the University of Illinois happened by. She had worked in my lab there, and we had dated briefly before I'd moved to College Station.

"Mmmmm," she said. "Mmmmm .... Harrison .... mmmmm ...."

She was telling me something about Dr. Harrison, but I only heard faint sounds. I tried saying, "I can't talk right now," but the feedback from my own voice was only noise. The more I tried making sense the more I messed up.

She finally just walked away, which seems a natural reaction to a babbler. I hadn't seen her in 14 years, and she was attractive and someone I really would have enjoyed chatting with. But .... "Hey, wait!" I could only think .... and there she went, and I couldn't do a thing about it.

So I wrapped up the poster and headed back for the streets. This time a big group was crossing the main intersection, so I herded in with the others.

Once back at the de la Poste, I saw my good friend and roommate at the hotel, Walter Bottje, sunning by the pool. I hurried past and headed for the room, where a maid was changing the towels in the bathroom.

I quickly plopped down on the bed and turned my back to the maid and faked sleep, so she wouldn't say anything. I had lost the language, but my senses told me what to do.

"Ted," Walter said when he walked into the room. "Ted?"

I didn't budge. Dr. Harrison then called to see if we'd like to grab a bite to eat, but Walter told him he'd let me sleep.

I lay in bed terrified by what was happening, too scared and even stubborn to try and do anything about it. I fell into a fitful few hours of sleep about 1 a.m., woke up at 5 and around 6, Walter got up and climbed in the shower. That morning, the same muddled state enveloped me, and the day proved one of my most terrifying ever.

"Trust God," I prayed. "Trust God." But deep down I felt pure terror at my helplessness, and the words I muttered to Walter that morning were anything but holy.

Research has shown that swearing likely comes from the right side of the brain, and my left side – from where most of analysis originates – had been damaged. Studies also have shown that patients with the entire left hemisphere of their brain surgically removed can still swear quite clearly. And I was using the right side of my brain wholeheartedly at that point.

"Ted, Ted, what's happening?" Walter asked. He found Brooke's number by calling College Station information and arranged for her to meet me at Easterwood Airport in College Station.

While packing my travel gear I tried looping my pants over the hanger, only to see them fall to the floor each time. I didn't know they were supposed to run through the hanger. And I cussed the entire time.

I finally just threw the pants in a bag while Walter retrieved the rental car. I didn't even know how to hang up a pair of trousers, and I was completely petrified.

I hadn't showered and looked ratty in a wrinkled T-shirt and a pair of jeans. Despite not knowing how to hang up a pair of pants anymore, I had no problem tying my shoes. This certainly had to do with the part of the brain affected by the stroke.

As Walter drove to the airport I muttered a few more vulgarities, and at the ticket counter Walter talked for me. The flight was near takeoff and so we cut through the line at the metal detector, prompting the security lady to holler. But considering my current state, she couldn't match me in the name-calling department.

OTHER VOICES
Friend and Colleague, Walter Bottje

I've known Ted since I was an undergraduate student at Eastern Illinois University in the Zoology department. He was a teaching assistant in a class I took. We worked in the same laboratory in graduate school at the University of Illinois and we lived together for a couple of years before he accepted the position at Texas A&M.

We kept in touch, and over the years we shared a room at professional meetings. Often, one of us would want to get caught up on some sleep and go to bed early.

When I returned to the room late the afternoon of Ted's stroke, he was in bed. I figured he was just napping, so I quietly left for dinner. When I returned about 11 p.m., Ted apparently was asleep so I went straight to bed.

The next morning, Ted was still in bed when I got up to take a shower. When I climbed out, he was sitting up.

"Good morning," I said. "You must have been wiped out last night."

Ted only shrugged. I asked if he was OK, and he shrugged again.

I realized something was wrong. Ted wasn't talking much, but when he did what came out was garbled – except for several four-letter words, which burst out extremely clear. Physically, there was absolutely no indication anything had happened.

Thus started an intense game of charades, as Ted acted out things to try and get a point across. Later, he'd say the main thing he wanted was for me to reach Brooke. As he got frustrated and couldn't get his point across, the vulgarities burst forth.

At one point Ted wrote three lone numbers on a piece of paper and pointed at the phone.

"Do you want to make a phone call?" I said. He nodded his head.

"Do you want to talk to someone in College Station?" I said. Again, he nodded his head.

"Is it someone in poultry science?" Shake of the head no.

"Is it a friend?" A nod yes.

I figured out he wanted me to call his girlfriend. I called information in College Station and got Ted's home phone number. Brooke answered and I explained that something was wrong with Ted. I told her some of the details and she said it sounded like the same thing that had happened to Ted six months prior.

Then, Ted was teaching a class and suddenly got dizzy and couldn't talk. The doctor diagnosed this as stress, and after a short time Ted seemed fine. In retrospect, it's apparent the earlier event was a mild stroke, but everyone – including his physician – missed the symptoms.

Brooke suggested he should come home immediately.

"Is that what you want to do?" I asked Ted. He nodded.

While I was out borrowing a rental car from another professor, Ted packed his bags and got ready to go.

At the New Orleans airport, I led Ted to the ticket counter and made the flight changes. During this time, Ted was pointing at the ticket and shrugging his shoulders. Later, he explained he was trying to tell me he didn't know how he was going to change planes at the Houston airport.

As Ted got on the plane, I flagged a woman at the ticket counter and explained I wasn't sure what had happened, but that Ted might need help changing planes in Houston.

Back at the FASEB meeting, the thought of Ted's incoherence distracted me, and I called that evening to see if he arrived OK.

Brooke answered and it was apparent she'd been crying and something was terribly wrong. Doctors said Ted had suffered a major stroke and that he wouldn't teach again.

A cold numbness and helplessness swept over me.

TED ODOM

Once I was on the plane from New Orleans to Houston, there weren't many passengers and I curled up in my seat. At Houston's Intercontinental Airport I knew my means of getting home was the

Continental Express terminal. I had tunnel vision and still could only see out of my right eye.

I couldn't stop and say, "Can you help me?" Simply because I couldn't talk.

I had been to Intercontinental enough, though, to know how to wander the maze to Continental Express, though I couldn't read the signs. I already had my boarding pass so I didn't need to say anything. At the gate I plopped down and studied the digital list of departures and arrivals, but I certainly didn't know when the College Station flight was leaving.

I was lucky it was a town named with two words to distinguish it from the others, as dizziness and vertigo distorted my thinking.

At the gate, the town with the two words finally moved to the top of the board, and I quickly got on the plane. There weren't many people onboard, so to this day I don't know if I sat in the right seat. I didn't say anything to anyone – I couldn't even if I had wanted to – I just turned toward the window and closed my eyes.

Brooke was waiting at Easterwood Airport when I finally reached College Station.

"Ted, what's wrong?" she said with a confused expression. I couldn't reply coherently and began crying, a show of gratitude that she was there for me.

We threw the bags in her trunk and drove to St. Joseph Hospital, where a physician – my regular doctor was off that day – checked me over.

The doctor put me in a room for the night under the suspicion that I'd suffered a serious stroke, although I had no idea at the time. I vainly hoped this all was resulting from a minor heart problem and everything would clear up shortly.

The entire time I was shaking my head that I didn't want to be admitted to the hospital, because I refused to believe anything bad had happened. I underwent a cat scan that night, and in the morning my doctor dangled keys a foot from my face. I reached out to grab the keys, infant-like.

"No, no," he said. "What are these?"

I perfectly understood what they were, but telling him was another problem.

"Of course they're keys," I thought. "Can't you understand?"

But I had no words to express my disbelief that he didn't know they were keys.

I also couldn't write, but I could draw figures. In a curious reaction to the stroke, I'd climb out of my bed and draw on a tablet against the wall. The images of the poster on the partition likely caused this, as those were the primary images I hastily scribbled on the tablets. I even reconstructed a chart showing the pressure on chickens' hearts that was a part of my poster. It's amazing how the brain works, but I wasn't a very good artist in such a sorry state.

In the days following the stroke I tried relearning the names in my family, and as Brooke wrote them down I'd practice right there in the hospital. That was my first lesson.

Pastor Malcolm Bane of College Station Baptist Church visited me in my second day at St. Joseph, and I wanted desperately to show I understood his coming. So I drew a "J" and then and a couple of "S's." I had thought a lot about God and Jesus since the stroke.

I tried writing, "Jesus" but it came out "JESU." Dr. Bane knowingly nodded his head and then he and my mother prayed for me. I prayed, too, but only in my mind.

Doctors released me from the hospital on the following Saturday afternoon following a five-day stay, and my home was a welcome and familiar comfort. I still didn't know I'd suffered a stroke, because my family and friends were protecting me to soften the blow.

"Boy," I thought, "this is some kind of heart problem." I naively figured I would shake all of the cloudiness loose and get on with life. Sure enough, I did get on with life – more abundantly than ever.

CHAPTER 8
Medically Speaking

What happened in New Orleans? Simply, blood pooling in my heart shot clots into my brain like tiny bullets. A brain scan showed a cloudy mass covering nearly a sixth of my brain, which is permanent damage from the stroke.

A stroke – or "brain attack" – occurs when a blood clot blocks a blood vessel or artery, or when a blood vessel breaks, interrupting the blood flow to an area of the brain.

The National Stroke Association defines a stroke as such: "A sudden disruption of the blood supply to a part of the brain, which, in turn, disrupts the body function controlled by that brain area."

A stroke kills cells in that immediate part of the brain because those cells lack oxygen and nutrients from the interruption of blood flow. This can also spark a chain reaction among other cells that didn't have their blood completely shut off. These cells are in a "state of shock" and can either live or die depending on what happens in the following minutes and hours.

When brain cells die, the body loses control of the area the brain once directed. This may include loss of speech, movement and memory. According to a March 2004 Newsweek cover article titled The War on Strokes, "Unlike a heart attack, whose pain generally sends people to the hospital immediately, strokes can be subtle, especially if they strike the parts of the brain dealing with memory or cognition rather than movement or speech."

A minor stroke may mean only minor side affects, while a larger stroke may mean paralysis or coma. How serious was my stroke? Here's part of an evaluation from an internal medicine specialist two years following the attack:

"Dr. Odom suffered a severe stroke due to hypertrophic cardiomyopathy. Due to the stroke he had severe speech deficit. He was otherwise a very functional and productive person. Due to subsequent speech therapy, his speech has improved quite a bit and he is able to communicate reasonably well at the present time. This is all due to the intensive speech therapy that he undertook.

".... (Further speech therapy is needed) as he is a researcher at Texas A&M and has to present important scientific papers and (at) international seminars from time to time."

Hypertrophic cardiomyopathy is defined by a thickening of the heart muscle, a swelling that can obstruct the bloodflow in and out of the heart. Cardiologists estimate that HCM results in two-thirds of sudden cardiac failures.

Here also is part of a letter from my speech therapist, Dr. Jean Foster, urging payment from an insurance company for further speech therapy following two years of recovery:

"Dr. Odom, at age 41, suffered a massive CVA (cerebral vascular attack) resulting in total loss of speech-language abilities. Essentially this stroke destroyed an enormous amount of brain tissue, primarily in the language areas.

".... Dr. Odom's recovery has been remarkable. I have been a speech-language pathologist for 25 years, and I've never had a patient with such massive loss make such a complete recovery. But it took hundreds and hundreds of hours of unbelievably hard work on the part of Dr. Odom, both in the treatment sessions and outside. Recovery was consistent. Speech-language is enormously complex.

"How is this any different from a patient who loses his hands in an accident, needs artificial limbs and must learn how to use them? Certainly there is medical necessity for that treatment prescribed. Dr. Odom lost part of his brain, needed an entire new set of speech-language rules and needed to learn how to use them."

This book's title is based on just such an analogy.

Amputees are known to experience a limb where there is none. Samuel Weiss of New York University writes that such "individuals have been unwilling to compromise with reality on an unconscious, primitive level. They insist upon retaining both the phantom sensations and the original extension of the absent limb. .... They insist on maintaining their former self-image despite the image facing them."

So went my reaction to a massive stroke, only my aversion to reality dealt with a mental loss, not a physical one. I compared my mind to a computer that had lost its program. And a computer without a program is nothing.

Aphasia, the inability to remember words, is like dialing a phone line in search of an answer, only to get a busy signal. A "Boston Aphasia Severity Rating Scale" that Jean Foster administered to me not long after the stroke showed a rating of "1," meaning "all communication is through fragmentary expression; there is a great need for inference, questioning and guessing by the listener. The range of information that can be exchanged is limited, and the listener carries the burden of communication."

Two years later Jean wrote I'd increased the rating to a "4" out of a possible five on the aphasia severity scale, defined as, "Some obvious loss of fluency in speech or facility of comprehension, without significant limitation on ideas expressed or form of expression."

According to the American Stroke Association, a person in the United States suffers a stroke every 53 seconds. Strokes afflict about 700,000 Americans every year, with about 75 percent surviving the initial attack. Of those, "nine in 10 will have long-term impairment of movement, sensation, memory or reasoning, ranging from slight to devastating," according to the 2004 Newsweek article. Strokes are the nation's third-leading killer and the leading cause of long-term disability.

The stroke association states that warning signs include numbness or weakness in the face, arm or leg (especially on one side of the body), sudden confusion, trouble speaking or understanding someone, trouble walking, imbalance, dizziness, sudden inability to see out of one or both eyes and sudden severe headaches.

Helpful websites in learning more about stroke include those of the American Stroke Association (strokeassociation.org), the American Stroke Foundation (americanstroke.org) and the National Stroke Association (stroke.org).

CHAPTER 9
A Dear Friend Visits

OTHER VOICES
Friend and Colleague, Walter Bottje

The day after I'd loaded Ted on the plane in New Orleans I decided to get to College Station as quickly as possible. I flew from New Orleans to Dallas and rented a car to drive to College Station.

At the same time a standoff between federal agents and the religious cult the Branch Davidians of Waco, Texas – an event taking place only several miles from the interstate I sped along – flooded the radio news in my rental car.

When I peeked in Ted's hospital room his face brightened. He tried talking but his frustration at his inability to do so once again led to the startlingly clear four-letter words.

At least one thing was evident: Ted already was trying to put the pieces of his life back together, and trying to figure out as much as he could about what had happened in New Orleans.

Brooke was exhausted, and so I told her I'd stay with Ted that night, Ted's second in the hospital. Every 15 to 30 minutes, Ted would sit up, point in the air like he had an idea and begin to draw on paper. One time he penciled a large rectangle with several small rectangles on it. Some small bar charts were then scrawled on the small rectangles.

I recognized these symbols as having been on his poster in New Orleans.

"Is that your poster, Ted?" I asked in the otherwise still night. He nodded emphatically.

Then he began to act as if he was in front of the poster chatting with people, and then he faked as if he didn't know where he was or what was happening. I realized he was telling me something terrible had happened at his poster.

These intense sessions lasted from 10 to 15 minutes, and he understandably became frustrated with himself or me when he couldn't get a point across.

There was a brightly-lit atrium in the hospital, and that first afternoon I suggested we visit it. So Ted put on a striped hospital robe and we started down the hall, but about halfway he abruptly stopped and pointed his finger in the air as if he'd forgotten something.

I started to go back with him but it was obvious he wanted to do this on his own. He disappeared back down the hall and after a few minutes I thought he might have been lost.

About then, he appeared around the corner in his robe and slippers and black sunglasses. This was a typical "Ted" thing to do. The sun was bright and he wanted to look cool in his shades.

The sun basked us in its warmth and I encouraged Ted to lean back and relax, because it felt good. This lasted about a minute before Ted popped up and began drawing on a tablet again. This time he wrote the number 10 and he pointed to it over and over again.

Then he scrawled a capital "S" and then an "e" and once again pointed repeatedly. Finally he began to draw a building with a large antenna on top.

"Is that the Sears Tower in Chicago, Ted?" I asked. He excitedly nodded yes, and then once again pointed at the 10.

I didn't understand the relationship between the two and could only shrug helplessly, as he continued trying through his motions and drawings to explain his thought. We then closed our eyes and rested a while. Later that day, Brooke returned to the hospital and I showed her the piece of paper.

"Oh, Ted is having $10 a month drawn on his checking account to pay off a bill to Sears," she said.

This was really sad and a little strange because Ted was only trying to tell me he was paying $10 to Sears, and it took him 30 minutes to partly get such a minute detail across to me.

I spent that Wednesday night in Ted's hospital room, and a physician or nurse periodically checked on him. Between these checks Ted always tried to get some new point across, so we didn't get much sleep.

Finally, some time early that morning we both nodded off, absolutely exhausted. Ted had done the same thing the night before with Brooke.

The next thing I knew the door flew open and Ted's doctor rushed in, banging on the end of the bed in the twilight of the dawn. I had no idea where I was, and it took me a while to get my bearings. The doctor began asking questions in rapid succession – he later said to see Ted's reactions – and Ted didn't understand and couldn't answer.

Then the physician shook his head and began talking to me, telling me Ted was extremely disoriented. The entire time I thought, "I'm pretty disoriented, too, as you would be if you had just fallen asleep and someone came banging into your room!"

"Ted's got some brain damage, and he's probably going to be institutionalized," the doctor said matter-of-factly. This I understood loud and clear, as I'm certain Ted did.

"I've known other cases like this, and those people never recovered," the doctor continued. "And you and anyone close to him should never think or delude yourselves into thinking that Ted can hope to recover. These other patients have been institutionalized and are now quite happy in their surroundings."

I was absolutely stunned. The doctor was talking about one of my best friends, a vibrant guy who just days earlier had been very active in teaching, research and in life. I also felt for Ted, because he just sat by as the doctor said this.

I thought, "How could this guy go on about Ted being institutionalized, or talking about someone being a vegetable, with him sitting right here?"

Thereafter each time the doctor visited he would talk to Brooke or me about Ted. Each time I thought he should have talked directly to Ted, but instead his attention only focused on others. It was as if Ted wasn't even in the room.

A couple of days later I told Ted goodbye. The drive back to Dallas where I would catch a plane back home to Fayetteville, Ark., was one of the saddest of my life. The sun was bright in a cloudless sky, and just north of College Station I drove through miles and miles of bluebonnets.

I prayed Ted would soon once again enjoy these simple things of life, and I also made a commitment to spend more time with my son, who was then 6 years old, so we, too, could enjoy the simple things. I reviewed my own life while thinking about Ted, and decided it was time for a few changes.

I wanted to believe what the doctors said wasn't true. But even with that faint optimism I'd have never believed he'd make a near total recovery.

Ted's stroke served as a wake-up call. All of us tend to be self-absorbed with our day-to-day problems and challenges, but at least for a while, many of his friends took a long look in the mirror and appreciated the simpler things. We realized how quickly things can change in life, and we took steps to make things just a little different.

CHAPTER 10
Family Ties

OTHER VOICES
Father, Ed Odom

Ted is a product of two families that came from a strong lineage of hard workers and hard times. My parents were farmers in Oklahoma, and my wife's parents did the same in Indiana. About the time my future wife, Toy, and I were born, the Great Depression had engulfed America, and within a couple of years the Great Dust Bowl swallowed our part of the country in the 1930s.

Our family was like many others in Vinson, Okla., as even the little children became part of the workforce. My earliest memories include wearing a sack for collecting cotton. Our family became part of the "Grapes of Wrath" trek to California in search of a better life recounted by author John Steinbeck in 1939.

Ted's mother's parents, Ora and Etta Weddle, also migrated to California from the Midwest, and my wife, Toy, spent much of her young life working in the grape industry in Van Nuys, Calif. Her father became a minister in the Brethren Church. Her mother worked in the food service at a Veteran's Hospital following World War II.

After Toy's father suffered what doctors then called a nervous breakdown – which most likely was a mini-stroke he had at the age of 35 – Toy's mother took care of her husband and their two young daughters while still working at the hospital.

Because of such adverse backgrounds, Ted observed a strong work ethic prevalent on both sides of the family, and adopted such habits in his own life. Without that work ethic ingrained in Ted we don't believe he'd have recovered to the point that he has. It never occurred to him to give up or give in. It was part of his tradition that you overcome adversity with all of your might.

Young Ted with his still-growing family

Mother, Toy Odom

Ed woke me from a nap on the afternoon of March 30, 1993, with startling news.

"Ted's had a stroke, but he's all right," he said. "He's not paralyzed but he's lost his speech."

By saying Ted was "all right," Ed only meant to calm my fears and let me know Ted wasn't dying. Ed didn't minimize the stroke's damage, because he told me Ted couldn't talk, although admittedly I didn't realize the severity of it all. I didn't panic and intuitively thought Ted would be OK.

My father, Ora Weddle, had died of a massive stroke at the age of 59 in 1961. But when he was about 35 he suffered the first of several mini-strokes, which at the time were labeled nervous breakdowns. Daddy didn't speak for a week after that first mini-stroke, and upon hearing Ed's news about Ted that immediately played upon my mind.

"He'll be like Daddy," I thought. "He'll speak in a week."

Ted smiled when we walked into his hospital room after driving straight from St. Louis along with our son, Pat. I was surprised at how well Ted looked. But his mannerisms were that of a little boy. At age 41 he looked like a kid all over again through those innocent, curious facial expressions that asked only one question: "What's going on?"

In retrospect, in those first few weeks of his new life, we never sat down and told Ted he'd suffered a stroke. We just supposed he knew.

Young Ted enjoys an Easter gathering with his family.

Brother, Pat Odom

Ted was the last person I figured anything would happen to because he's always been such a health and fitness nut.

I was very emotional when my father called about Ted because he represented perfection. He seemingly had everything going right in his life and excelled at all that he did.

When we arrived in his hospital room, Ted looked normal physically – I was expecting him to look like he'd been hit by a truck or something akin – although his eyes seemed slightly dazed.

The big shock came when he tried speaking. About the only word he muttered that first day was, "How?" as a response to anything said. But I could also tell that Ted's former self was hiding in there, as he tried in vain to tell us specifics, only the words wouldn't spill forth.

In the ensuing days Ted's first words to me were only pronouns like he, she, him and her. "Him, him, him" would burst from his lips and you could see the frustration at his inability to express himself.

Ted and I were close while growing up, so we had many past references, like our love for the cartoon "Droopy Dog," that no one else understood. Despite Ted's condition we could still make allusions to our past. We felt like we were communicating, even when we weren't actually talking.

Maybe the most remarkable thing through all of this was Ted's ability to still possess a sense of humor – even if that only meant trying to imitate Droopy. We watched lots of television as children, and we often made references to shows and movies we'd seen together. So for Ted to try and imitate Droopy in those early days following the stroke was a good sign, because he already was connecting with his past.

Through his entire recovery from the first day I saw him, Ted has had the ability to laugh. Often comedians or generally funny people use humor to help cope with an extremely painful thing in their lives. To his credit, Ted has done the same thing.

Ted could have acted depressed or felt sorry for himself, but instead he kept that infamous humor intact. I couldn't figure out how someone who'd just suffered so much damage to his brain had anything to laugh about, but to this day he's maintained that sense of humor.

If Ted hadn't been so dynamic before, I'm not sure he'd have recovered to the point he has, because that same determination and humor has kept him afloat in the tough times.

Father, Ed Odom

When the doctor called to first tell us about Ted's stroke, he said they were deeply concerned because there was no apparent cause for it. Ted didn't smoke and he didn't drink excessively, so how were they going to prevent the next attack if doctors didn't know what they were treating?

In the meantime it almost seemed worse than death for Ted to have that brilliant mind trapped in his body, because of his intense personality and need to constantly communicate and achieve.

Ted with sisters Cathy and Lauri at Christmas time

"That's only reflexes," doctors said of Ted's early repeating of the word, "How." The more the doctors talked the more upset we became, but such negativity made us more determined to show them Ted would prevail. Our first thought was to bring him back to our home in St. Louis, but his happiness and familiarity was in College Station. The last thing he needed was a world void of those things.

"Take him to St. Louis with you, give him an aspirin a day and don't bother with speech therapy, because that's useless for him," one doctor said.

Such talk only served to raise the fighting spirit in us, and we became determined to find a path to recovery. We weren't going to write off Ted as a lost cause, even if that meant he needed to stay in College Station.

Leaving him was probably the hardest thing we've ever done. We thought if his relationship with Brooke suddenly ended, he would pretty much be alone.

Saying goodbye on any occasion, even under good circumstances, has a sad side that can lead to depression and homesickness. But this was different, almost like leaving a little boy alone in the world. The hard part as parents was the helplessness in witnessing Ted's struggle. To a degree we just had to go blindly and believe the Lord would look after him.

You hope that as you've grown older your spiritual strength deepens and you know there are some things out of your hands. You say it's up to the Lord.

You pray that strength will flow to your son and he'll have the courage and belief that he can move forward and overcome.

Mother, Toy Odom

We had to let Ted try and get well in College Station. If he had lost all of this familiarity, then he'd lost all of his life.

The reality of this new beginning for our son hit hardest when we got home and I realized we wouldn't be talking to Ted on the phone anymore – at least not for a while. Before the stroke we would talk for hours, and oh, how I loved those calls.

I was always taught to pray about matters, and that if you wrote them down they were then in the Lord's hands. You write them down and let the Lord take care of them, such as I did when we left Ted in College Station.

CHAPTER 11
Hospital Daze

Following my release from St. Joseph Hospital in Bryan five days after the stroke, the doctors' only orders were to take an aspirin a day.

I still didn't know I'd suffered a stroke, but family and friends engaged in hushed discussions in the kitchen and on the phone, and I knew something was up. They also would talk very slowly to me.

"There's - another - doctor - who's - going - to - check - you - out - and - see - if - he - can - help - you," a one-sided conversation would go. And then that loved one, trying so hard to make me understand, would carefully repeat the line.

It was a bit like talking to a foreigner who understands little English – like if you spoke slowly enough he'd finally understand. Now I knew how those A&M graduate students from China and Taiwan felt when we'd speak slowly to them in English.

They'd smile even when they didn't comprehend what you were saying, especially if you had a friendly face. And now I was the grinning hyena. I'd just shake my head and smile widely, though I hadn't a clue what was going on.

My parents, my brother, Pat, and Brooke were keeping a close watch on me the entire time – and meanwhile they were calling around for doctors who might best understand what had happened to me. I couldn't understand how a heart problem, which I still believed I had, could impair my speech.

When I first arrived at my house from the hospital, I hurried into my room and flopped on my bed, like a kid on his first trip home from college who was so glad to be back in deeply familiar environs.

Kato, my 108-pound Rottweiler, rushed into the room to greet me, and he rested his large paws around my neck. He had never done that before, and it was one of the most comforting things that had happened to me following the stroke. I'd owned Kato for six years – since he was seven weeks old – and a dog's undying loyalty is one of life's true simple treasures.

In trying to reintroduce me to familiar things, Brooke and Pat also drove me the couple of miles to my office the day after I got home. It seemed a fast way to reconnect with what I'd had before, but seeing a stack of paperwork on my desk seemed a bit overwhelming.

"How will I ever get through all of this," I naively thought, as if that was still so important in light of what faced me now.

A day later, we headed south for Houston's Hermann Hospital, where my family had gotten in touch with a renowned cardiologist through my friend and colleague, Bill Chilian. There, I received some get-well wishes from friends and family, including from my youngest sister, Sandra. Her card first informed me of the stroke.

I couldn't read the note inside, but the word "stroke" somehow stood out when a family member read the card to me.

"Hello dear brother," it read, "I'm so very sorry to hear the news of your stroke."

The startlingly severe word echoed as I deciphered its meaning, and suddenly I knew this was more than a minor heart problem.

***

Houston is only a little more than an hour across the mostly-flat prairie from College Station, but in my predicament it might as well have been across the Gulf of Mexico, because I seemed so terribly far from home.

That first evening, as doctors and nurses subjected me to test after test, the skies slowly blackened as night fell through the windows that looked out on beautiful Hermann Park, which disappeared into the darkness.

I'd always figured doctoring to be a 9-to-5 job unless there was an emergency, but here they didn't seem to stop.

"These guys are working late," I thought. "When do they sleep?"

I thought in images if not really words – you just knew the activity without the language attached.

Doctors peppered my arm with needle marks, and every two hours a nurse checked my temperature, heart rate and blood pressure.

Still somewhat confused about what exactly was going on, despite the introduction of the word "stroke" into the chaos, I felt a slow disconnection from what little had seemed left of my prior life. I knew I wasn't alone, but I was deeply scared – a prisoner in my own body that I no longer really knew.

A hospital, despite its necessity as a place to heal the sick, in many ways can seem like a prison, because the only way out is to escape, if you haven't received permission to leave.

Such a thought caused me to peer out the windows many hours my two weeks in Houston – I can't imagine what a life-sentence in prison must do to a soul – and watch kids and families play in the park below.

"Ahh, freedom," I'd murmur to myself.

Which really meant that I only wanted to go back to work, because that had been my prior identity. I had never really visited people in the hospital because it seemed a little scary. Kind of like prison.

Possibly the most important thing I learned during my two weeks in Houston at Hermann Hospital is how much it warms a soul to have visitors during such a confusing, disturbing, lonely time.

Before, I had little time for those who couldn't keep up with me. But in the hospital, I suddenly realized with alarming reality how important it is to take the time to care – and show you care.

It's simply extremely important to visit someone in need, and doing such – even if that person is a stranger – can have immeasurable consequences. Prayer together, too, is extremely powerful.

Many of the elderly don't have a lot of friends because so many of those they've been closest to have died, and it seems often younger people aren't comfortable visiting with older folks. Maybe it's because they remind them too much of their own mortality.

One night a silhouette appeared in the doorway of my hospital room, and I immediately recognized the kind, familiar face of my psychologist from back home.

"I'm sorry," he softly said. "I'm sorry I didn't recognize you had a problem that was more than just stress."

I deeply appreciated his coming, and tried hard to communicate that it was OK, and that a proper medical diagnosis wasn't his responsibility, anyway. We both felt better after his visit.

Honestly, many of my friends never visited or called in the weeks following my stroke, possibly because of denial or fear that someone so young – one of their peers, for goodness' sakes – could suffer such a life-altering affliction. At the age of 41, I became disconnected with several of my previously close friends because of the stroke.

Prior to my comeuppance the only thing I really knew, or cared, about strokes was that old people have them. Or so I thought.

***

After conferring with nurses, my father said it would be nice if I attended an Easter service across the drive from Hermann Hospital. That Easter, in fact, was the first I didn't focus on dyed eggs as an integral part of the day.

Doctors agreed to let me walk across the drive with Brooke and my dad, along with the ever-present IV on wheels that thinned my blood.

By all accounts it was a nice service, but I couldn't understand the minister and when I tried singing, the words seemed all out of order. No one was going to mistake me for a member of the choir, that's for certain.

But a funny thing happened in that moment of utter isolation – so close to my father and girlfriend yet so disconnected from reality – a moment I still keenly recall for its purity and love.

In that instant of solitude, I truly and deeply comprehended the meaning of the Resurrection, and I became giddy with joy.

Truly appreciating the Resurrection and the immaculate truth that Jesus Christ died on the cross for our sins was such a mind-numbing and humbling thought on this Easter, I absolutely realized I'd been given more years on this earth for a reason. And just as my new personality dictated to start over, I also was childlike in my newfound reverence for the Lord as my savior.

Jesus summed up such a new beginning perfectly in Matthew 18, verse 3: "I assure you, unless you turn from your sins and become as little children, you will never get into the Kingdom of Heaven."

Granted, I didn't know much about the road ahead, or even what had afflicted me in the first place, but I wholeheartedly realized by the end of that consequential service that I was in good hands – God's hands.

OTHER VOICES
Mother, Toy Odom

When Ted was 10, I read a little meditation in Sunday School about Easter and its meaning, and none of the other kids in the group seemed very interested. But Ted eagerly absorbed the story, and he wanted to hear more of the Scripture about Jesus' Resurrection. Ted always had that spiritual side, no matter how suppressed at one point it became in his life, something that proved vital to him in his recovery.

TED ODOM

Following the Easter Sunday that served as my own personal resurrection through Christ, the nurses back in Hermann Hospital offered up several injections of tiny doses of hell.

"We're going to inject dye eight times," one nurse cautioned, "and it might hurt."

I soon figured out that was a light description. The injection, meant to discover any possible lingering blood clots in the brain or in the vessels, made my brain feel like it was melting. Mam-ma used to talk about how hot she presumed hell to be, and she used a startling example to get her point across.

"If you're in hell," she'd say, wagging a worn, wonderful finger before me, "it's not even close to what it feels like to burn your finger."

And I knew, through my own youthful antics, that was plenty painful in its own right. This angiogram wasn't hot as hell, but it was the hottest thing I've ever felt. Thankfully, though, the test results showed no clots.

"You're clean," the doctor said.

Such news helped ease any lingering depression in the weeks following the stroke, although by the second week in the hospital I was like a caged tiger, ready to bust through the window and gallop home to College Station. I'm fortunate I had such good, attentive care from the doctors and nurses in those final few days at Hermann, otherwise I might have suffered a stress-related setback.

I'm also eternally grateful to Brooke for her help and understanding during that period, because I couldn't even bathe myself. Brooke would bathe me, and sitting there in that white tub I didn't even realize I should be ashamed I was naked – much like the innocence the biblical Adam possessed before biting into the forbidden fruit. Only years later, and long after we'd broken up, did I realize how much Brooke took care of and looked after me.

***

A kind and gentle speech therapist, Ann Sharker, began working with me on my third day in Houston. Ann used flashcards with pictures to begin my relearning of English, and I felt bad that I was having so much trouble comprehending something that otherwise should be so simple.

One picture she held up and asked me to explain caused utter confusion, because all of the activity in the photo made my mind blur.

The picture showed a house, a lake with small boats on it, a sidewalk winding around the lake, kids on the sidewalk around the lake, and cyclists passing the kids on the sidewalk around the lake. See what I mean?

"I'm overwhelmed!" I thought, shaking my head no at this image.

I didn't even know where to start in describing it. A picture with just a sidewalk would have been perfectly fine with me. I needed learning in very small measures – half a teaspoon at a time – in those few weeks following the stroke. So, I would cover one part of the picture with my hand and absorb it a little at a time.

In the morning, my dad would read the Houston Chronicle to me. I'd glance at the paper, but all I'd see was a jumble of black newsprint. Billboards, too, so colorful out the windows, may as well have been scribed in Greek. I was so childlike in that sense of trying to comprehend all of those things written around me, and deep down I knew I'd have to start from the beginning if I were to function in society again.

One of the primary problems, I soon discovered in relearning the language, is that so many English words have silent letters – but I would see and say all of them.

For example, light became "ligghuht." Try speaking that language, or better yet, try having someone understand it. I simply had no idea which letters to skip over.

My family and friends who visited me during those two weeks also spoke of an event unfolding about three hours to the northwest – the Branch Davidian situation in Waco.

On the Monday of my release, April 19, 1993, the Davidians' compound burned to the ground, killing more than 80 people inside. My family watched the TV in disbelief, but I could barely recollect the long, grim standoff in the days prior to the stroke and didn't know exactly what was going on.

OTHER VOICES
Father, Ed Odom

One of the remarkable things following Ted's stroke was that Bill Chilian – a friend and colleague of Ted – had conducted research with a renowned cardiologist in Houston.

Ted met with the cardiologist and a Houston neurologist his first day at Hermann. The neurologist later seemed incredulous as he discussed his findings from tests he'd run.

"Did you know your son had two minor strokes prior to the big one?" he asked. "The old damage should have been seen."

The doctors immediately connected us with a speech therapist. This reaction starkly contrasted with a couple of doctors in College Station who had said, "You'll be wasting your time if you try and get a speech therapist."

The thought that the doctors who gave Ted no chance of recovery were correct crept into my mind, but I didn't know how to tell Ted's mother or brother about my doubts. So I kept them to myself and had faith that God would take care of things.

One day late in his stay Ted drew stick figures of a boy and girl on a tablet that was constantly by his side.

"What on earth could that mean?" I thought before it dawned on me.

"Is that Dick and Jane?" I asked. Ted nodded excitedly.

"Do you want a Dick and Jane book?" I asked. He nodded again, this time even more profusely. I hurried to his brother.

"Pat, Pat, go to the bookstore and get some Dick & Jane books," I said.

Pat, as you can imagine, cast me an, "Are you out of your mind?" glance before heading out in search of the elementary reader. I believe that's where Ted's recovery really started, because at that moment he knew the severity of his situation.

Every day in Houston, Ted recovered a little more, enough to where we could tell by his grimaces that he really disliked the hospital food. As a child Ted had always enjoyed a good hot dog, in fact he considered that a decent dinner.

"Shall I go out and get hot dogs?" I asked Ted one morning, and he smiled. I quickly returned with a package that he wolfed down for breakfast.

You could tell by his grin that a few Oscar Meyer wieners sure beat skim milk, fruit and grainy cereal on this morning.

This was a sign he was becoming a little more aware of his surroundings, because he began recollecting what he liked and disliked. And we knew, through his mannerisms even if he couldn't tell us, that Ted didn't like being in the hospital this long. It was nearly time for him to go home.

CHAPTER 12
Back to School

I didn't truly know my name until I began working with Dr. Jean Foster, a wonderful speech therapist in Bryan, a few weeks after the March 29, 1993, stroke.

I'd returned from Hermann Hospital and began visiting Jean daily shortly thereafter. My first assignment was learning to write my name. I figured it must be right because everybody called me that, anyway.

In those first days following the stroke I'd have to draw a rudimentary figure to try and communicate or rub my tummy when I was hungry. That's also about how my first few sessions of rehabilitation with Jean went.

Jean's system was set up so the patient wouldn't be so overwhelmed that you can't even start on the drills. My research was still intact following the stroke, but I'd lost the tools to talk. You have to start somewhere in relearning the language, and that meant clawing from one word to two to a simple sentence to a paragraph and so on.

Early on if Jean were to say, "banana," I'd know it was a yellow fruit, although I couldn't pronounce the word myself. It was a bit like possessing a slight grasp of a foreign language.

Also, if I had an image before me instead of a jumble of words, well, that was much easier to work with. For instance, I easily could circle a picture of an apple or banana during an exercise, but had much more trouble with identifying the spelling of each.

I could only circle or underline answers at the beginning of rehab. Trying anything else swamped me. I messed up early and often in the drills comparable to those for a young elementary school student.

For example, one exercise called for me to cross a line from a list of nouns like birds and babies to verbs like cries and sings. I messed that one up. Babies can sing, I thought. But Jean wanted no part of such rationale. If I could have talked I'd have argued, "Hey, how old is the baby?"

In my mind, at least, I already was questioning statements like that, much like a curious first-grader who believes he's smarter than his classmates and must point out silly things like some babies can sing.

It was the repressed professor in me lurking among a badly swollen brain.

OTHER VOICES
Speech Therapist, Dr. Jean Foster

Ted never really had a singular, defining breakthrough moment during his sessions. His progress was more of a gradual process through the course of 16 months. Typically, as improvements become more and more complex, you notice them less.

It's much more dramatic if suddenly you say your first sentence. For instance, if Ted said, "I see the apple," after not being able to do anything other than point at the apple or just choose between an apple and a pear, that's pretty dramatic.

It's at least equally as difficult language-structure wise, however, to later figure out something like the proper word structure for combining a conjunction. For example: Where to put the "and" in "I want to do this and I want to do that."

The brain still must learn the same things it did the first time around, but such learning isn't as noticeable because after those first dramatic breakthroughs you're simply refining the language. Once you get the subject and verb in order, the growth involved in relearning the language isn't as pronounced – but it's just as important.

TED ODOM

Have you ever wanted to go back to when you were in kindergarten or the first grade, an age of innocence and newness, and have the chance to do things over? In a sense I got that chance, because part of the process of relearning the language and numbering meant remembering the techniques I used as a boy to memorize math and English.

And just like in elementary school, the drills grew a little harder as the days grew longer, like the one where I paired Charles Lindbergh with "pilot" and Lyndon Johnson with "president" in a list of options.

As I began putting vowels and consonants together, Jean gradually had me recite aloud all of my assignments, to help organize the language.

And she always timed me on the drills. She'd also have me listen to cassette tapes where the narrator slowly said a word and I'd repeat it. I took my time with the tapes, and I didn't care how long it was going to take me to learn those first 75 words.

Remember how as a youngster I'd be so impatient I'd slap together a model car before the paint had even dried? The stroke certainly taught me patience, beginning with that cassette tape. I'd practice the words relentlessly while listening on a Walkman and walking around for exercise, and that first tape of 75 simple words took me three weeks to complete.

"I don't know if you're ever going to get through with these," Jean said one day, half-joking, half-serious.

But when I had finished with it, it marked one of my early breakthroughs. Early on, those 75 words may as well have been 750 million, until I learned to take them one at a time – and not concentrate too much on the finish line.

Jean had me constantly work on phonetics, and I'd use Brooke's old makeup compact to watch how my mouth formed certain words. The phonetic spellings to me resembled the symbols seen on the old pyramids, and the phonics were about as easy to decipher as Arabic in the beginning.

As my learning progressed I'd practice consonants and vowels and names and address and days of the week. I had a lot of trouble pronouncing the months. It's much easier to say, "Monday," than "September," as the names of the months vary much more than the days of the week. If I would say, "January," for instance, I very slowly would break it into four distinct syllables: Jan-u-a-ry.

Drills also included addition and subtraction, and I picked up numbers faster than words. I had lost my language because of damage to the parietal area of the brain, while the portion of the brain set aside for numbers and math wasn't as severely damaged. As the swelling in the parietal area decreased, my thought process also began to clear.

I also realized early on that an object is easier to learn than action. A verb or any kind of word describing movement usually caused me confusion, because of its complexity.

Naturally a professor – particularly one who took such great pride in his previous accomplishments – should have been frustrated with such menial tasks as choosing between an apple and a pear, but I didn't know any better.

I only wanted to work until I had the right answer, and everything had an answer if you toiled long enough to discover it. Sometimes baby steps are the best way to reach a destination, no matter how long it takes you.

If I had perfect homework, Jean would reward me with gold stars, and I worked hard for those stars. Those little golden gifts – along with the occasional pat on the back – were small reminders I was progressing back to where I'd been before, at the front of the classroom.

While both Jean and I possessed our doctorates – in different fields, of course – our positions in my current position were clear: She was the teacher, I was the student.

Arguing with her for the sake of argument wouldn't have accomplished anything. Besides, I didn't really know anything.

***

The sessions were set up to have lots of homework, and I had lost the idea I needed to be at work, make a living and pay bills. As my homework load increased my relationship with Brooke fizzled. I didn't have any idea how to talk about relationships and iron out problems. Sometimes it felt like I only had enough energy to breathe, and I used whatever bit remained for my homework and study sessions.

Brooke had taken care of me like an older sister would, and I appreciate everything she did for me in my most trying times.

The intense study sessions aside, possibly the most important point in my recovery came when I understood that I'd suffered a stroke, and wasn't the man I once was.

One day two months after the stroke I walked into Brooke's office in the A&M veterinary school and a 5x7 framed picture caught my attention. In it, I was grand in an expensive suit and tie and possessed a determined, self-assured stare.

It was the same photo that hung in the halls of the Poultry Science Department along with the pictures of the other professors, and at that very moment I understood I no longer was the person in that photo, and that realization nearly knocked me to my knees.

I fell into a chair in Brooke's office and cried softly in my hands, as Brooke put her arm around me and began crying also.

A wave of pent anguish coursed through my every being and I was emotionally overwhelmed, no longer naive to my new state of mind. Suddenly I knew I'd been a professor, and I didn't know if I'd ever get back to that.

The only thing comparable emotionally to this feeling of loss was going through a divorce. Then, I'd run miles and miles alone, telling myself, "It's OK, Ted, it's OK." Inside I believed I was a loser because my marriage hadn't worked out.

Now I'd lost a wife and myself – or at least that confident guy in the picture. The best way I can describe it is this: what if you woke up tomorrow morning, a stranger to your skin, hands, face? And what if two months later, as you reacquainted yourself with yourself, that you figure out the lively, alert person now reading this is buried and gone for good?

It's an unsettling idea, one I lived through. The day after I broke down after glancing at that picture, I attended a session with Jean and immediately upon seeing all of the menial lessons spread before me, I cried once again.

OTHER VOICES
Speech Therapist, Dr. Jean Foster

Ted thought, as is normal, that if you worked really hard, eight weeks down the road all of this would be finished. But there comes a realization that it's a step-by-step procedure and there's no way the recovery is going to all happen overnight, no matter how hard you work.

There's an enormous amount of panic that goes along with that, until the brain essentially heals to the point where you can deal with it.

The better you become, the more you realize what it is you have to do – and what happened to you and where you have to get to. It's a real trauma to your system, just like any major illness would be.

TED ODOM

I hadn't grieved for the loss of myself until that point of seeing that picture, and then I grieved a lot. The vibrant man staring from that framed photo no longer lived, and although I'd enjoyed a rebirth of sorts, that didn't mean sometimes I didn't miss my former self-confidence, along with some old friends who no longer connected with me.

But through faith I persevered, although it was the toughest time of my life. It also marked the most important point in my recovery. At age 41 I was at a crossroads, and I could have given in to the temptation to just quit and accept my current state, or I could press on. It seemed every minute I was tired, and I only wanted to get back to the way things were, and skip the sweat and tears and frustration involved in the process.

But I also knew God was with me. With His help I simply got over it and lived on, and tougher sessions and lessons waited.

CHAPTER 13
In Foster's Care

One of the more pleasant side effects of the stroke was a renewed understanding of the relationship between the teacher and student, and that respect from both sides is vital to learning.

Jean commanded my respect, and returned it in kind. This was something I carried back into the classroom as a professor, because before I didn't bother trying to understand the supposed rotten apples who didn't seem to want to learn.

Sometimes people have different ways of learning, and one of mine was to place myself in the action during drills, no matter how short the sentence: Ted is running. I'd use that example because I'd been an athlete.

That was the best way I knew how to get back to my prior standing in life. When you lose everything, the more I used my name in sentences the more I discovered who I had been.

Simply saying Bob or Jane didn't help me. Learning how to form a sentence meant starting from the most basic connection of a subject and verb.

"Ted ran" eventually became, "Ted ran yesterday."

It's hard to appreciate the beauty of a simple worksheet with a few similar sentences scrawled on it, but I certainly can. I fought to learn about something as simple as a pronoun, and I'd typically work the harder lessons first – and feel drained of all energy at the end.

I also began reading everything aloud, as that was part of the learning. Before the stroke, words continuously flowed from my brain to my mouth, now I had to think about the language before using it to communicate.

A constant feedback piercing my brain, one that still exists on a certain level, also made for difficult conversation. Imagine trying to carry a conversation as you wear headphones blaring a screeching tune, and you get the idea.

I still wear cotton in my ears to help drown out distracting noises that keep me from concentrating on the task at hand. Cotton helps soften the surrounding voices and noises, because it's hard for me to focus when there are many activities going on around me.

I also still use a "focusing stone," which I rub in my right hand. It's very smooth and relieves a lot of stress, and, living up to its name, helps me to focus on the task at hand.

Despite the potholes in the road to recovery like the loud, disturbing feedback from my own voice, shades of my old self were beginning to emerge from the depths of the stroke a few weeks into my lessons with Jean. I'd giggle at sentenced that contained innuendos – real or imagined – meaning I was beginning to understand the language more than just on its surface.

"It's OK to sometimes clown around, Ted, but we're going to have to break you of that behavior if it continues," Jean said, trying to keep me focused on the lesson before me.

I was discovering humor all over again, which was a great feeling, even if I was the only one laughing.

OTHER VOICES
Dr. Jean Foster, Speech Therapist

The first time I ever saw Ted he was in the hospital at St. Joseph's, only a few days following his stroke. At the time I was contracted with the hospital to visit their stroke patients.

He hadn't really said anything up until that point. He was alert, and I got him to imitate me counting.

"One, two ..." I'd say.

"Three, four ..." he'd respond.

Those kinds of patterns tend to stay in your head, because they're almost automatic responses.

Although we didn't even get as far as 10, Ted was just thrilled, as that was the first thing he'd really been able to say. He also was able to point at items when I'd say their name, so he had some basic understanding of language. The first session was extremely short, which they usually are, about five or 10 minutes.

"His speech areas aren't totally gone," I thought as I walked out of the room. The doctors hadn't given his parents a lot of hope at that point, but I felt certain he would be a candidate for an attempt at rehabilitation.

After that, I didn't see Ted for a few weeks, as he was sent to Hermann Hospital in Houston.

When he came back to Bryan-College Station, his first session consisted of writing his name and then some simple words, although they consisted of many, many spelling errors.

Early on, it was extremely difficult to joke with Ted, because everything had become so literal and concrete in his world. He'd have no idea why someone might say, "Wow, you really hit the nail on the head," when there wasn't a nail or hammer in sight.

Ted visited with me almost daily over 16 months, and in that time his recovery was dramatic. He went from virtually having no verbal speech to being a functioning professor. That's amazing.

He relearned the language, and he did it from scratch – just like he was in the first grade all over again. And it certainly wasn't easy, as Ted put in hundreds and hundreds of hours into his recovery.

Ted was relatively young to have had that kind of damage, but he was smart and very determined. Being young has an awful lot to do with what your brain is capable of relearning, and 41 is young for a major stroke. There's a big difference between what a 40-year-old is capable of relearning compared to an 80-year-old.

Any exasperation on Ted's part toward the menial tasks involved in relearning the language was minor compared to the amount of determination that he had to do it. I would give him 10, 15 or 20 pages of homework from session to session to accomplish between, say, a Monday and Wednesday, and he would have it done. I don't think there was one time when he didn't have his homework done.

Early on it was Dick & Jane kind of things, and tasks like matching the word "cat" to a picture of a cat. He would always have those assignments done and then he would ask for more.

He sometimes had the feeling of, "Why am I having to do this?" but he realized it was part of the process, and he did it.

What I told Ted to do, he did. He was so bent on trying to relearn as much as he could as fast as he could that he really treated this like a pupil-teacher situation, and his having been a teacher might have helped that. You have a better understanding of that relationship when you've been on both sides.

Ted's big goal was to get back into his line of research and teaching, and he understood it wasn't going to be easy. He sometimes joked that I was mean, but we both knew that I was only mean in the sense that I stressed that the lone way you're going to get better is if we perform steps A to Z.

We spent a lot of time looking ahead and to what he was going to accomplish rather than saying, "This is really terrible and I can't do this." That's not to say there aren't peaks and valleys in such a process, and some asking of "Why me?" along the way. But you can't dwell on that for very long – and to Ted's credit, he didn't – or you don't get accomplished what you need to. Ted also was fortunate to have the type of job he did, because there are probably a lot of employers who in similar situations would have just said, "I'm sorry."

Probably the most difficult thing for Ted was going through such an emotional situation and not being able to communicate the things he was feeling. You have extremely deep feelings at that point, but your language is gone. That must be enormously frustrating, because you have all of those things inside of you and no way to let them out.

Ted's struggles and triumphs are inspirational especially in the sense of seeing how much he now cares about what other people are going through, something he never really thought about before because it didn't touch his life in any way. He now feels, because he's gotten his language and speech back, that he can show others it can be done for them as well.

And there's no better way to show others than to say, "Here I am. I can talk and I went back to work. It's going to take a lot of work and effort, but this is what can happen." There's nothing more inspirational than being a living example of such a feat.

People meet Ted and say, "Oh, he's not hurt that bad." I often wonder that if somebody just met him, if they would know anything had been wrong at all. I applaud him for writing a book on his recovery, because there are so many young stroke victims – meaning under the age of 60 – in need of an inspiring story because of having endured such a life-changing event.

You can't make promises on the extent of a recovery, but more and more research shows that the brain is capable of amazing things, even when having to relearn a language.

Though certain areas of Ted's brain were severely damaged, he was able to recover because of his perseverance and determination. Ted did more hours of at-home therapy than anyone I've treated. He just absorbed the information as fast as I could get it to him.

People need realistic hope, not false promises. But they also need to know that an enormous amount of work can pay off, even for a severe stroke victim.

CHAPTER 14
Slow Return of the Everyday Things

So how long before I grasped how to do the "everyday" things, the chores necessary for daily living? It all depended on what part of the brain the activity stemmed from.

Tying my shoes? No problem, I wore running shoes back from New Orleans the day following the stroke.

Hanging up pants? I had to relearn this process, because I couldn't understand how the pants slid through the hanger. I had to watch others do it first, and once I had that image implanted the method snapped back into my mind.

Tying a tie? This escaped me, and it took me nearly a year to master the procedure. On March 29, 1994 – a year to the day following the stroke – I wore the entire suit I had on that day in New Orleans to my session with Jean, right down to the socks. I did so to celebrate my comeback and what I considered my second birthday – and the fact that I could once again tie a tie.

Learning to do as much was one of the processes that I remembered more once the swelling in my brain receded. Until then, I had no idea. Brushing my teeth? This was a very familiar process and one that never escaped me – thank goodness for my friends and family.

Making breakfast? I carefully watched Brooke make coffee in the morning, and finally relearned how to do so myself. I emulated Brooke on many things, in fact, in the kitchen and elsewhere.

Using the bathroom? I never lost the ability to understand this process – thank goodness. This was a simple thing to remember, and more reflexive than anything. Cleaning house? These also were activities I knew how to do, and I clearly understood you've got to constantly clean the place to keep it livable.

Using the TV remote control? Ahh, some important things you just don't forget how to do – and this was one of them. With years of recovery ahead in other vital areas immediately following the stroke, the remote in the hospital still felt as comfortable in my hand as a big knife in Jim Bowie's.

Paying bills? Brooke paid my bills until I was comfortable signing my name – and figuring out exactly what the bills were for. This took months. Talking on the phone? This still can be difficult for me, because I rely heavily on visual cues and a person's mannerisms to try and understand what he or she is saying. But with patience on both ends, I can carry a conversation and catch up with family and friends via Ma Bell.

Driving? Because of the interaction with the other drivers, we naturally wanted to err on the side of caution in my return behind the wheel. Brooke toted me around for the first six weeks before the doctors allowed my climbing in the driver's seat, and then for a while she rode very attentively in the passenger's seat.

One of the biggest parts of driving, of course, is reading the streets signs – especially the speed limit signs. And since I'd had the least trouble with numbers from the beginning – because that's not where most of the damage to my brain occurred – "55" and "35" and such were decipherable.

But for the first time in my life, I really had to plan where I was going to drive – street by street, turn by turn. I was like an explorer in the wilderness because I plotted every move.

When I'd finish a session with Jean Foster that had sapped much of my energy, I'd concentrate hard and drive very carefully and slow, much like many of the elderly do.

Meanwhile the road signs were a little like the homework assignments of Jean – the more I saw the more recognizable they became and the better I did. I focused hard on driving and did everything by the book, because I didn't want the police to stop me.

I've been stopped twice by the police (at least up until this publishing) but haven't been ticketed. The first time came in 1995 – two years after the stroke – when I'd run up to the nearby Circle K convenience store to buy some vanilla ice cream around 10 p.m.

The College Station police officer had pulled me over because he said I wasn't careful enough pulling out into an intersection. I have trouble talking late at night, so the officer immediately suspected something was up.

"Have you been drinking?" he said.

I shook my head.

"Then why are you slurring your words?" he asked.

I then slowly explained that aphasia caused me to search for and sometimes stumble over words, especially late at night.

He smiled and said, "You better get on home. Wouldn't want your ice cream to melt on your seat."

At night, my energy has typically waned to the point that talking is a chore, not a habit. It reminded me of a favorite Star Trek episode where an alien could only communicate with his brethren through thoughts, and not language. He had extremely defined thoughts and didn't stumble over words.

But when that alien tried talking to the Star Trek crew, his words were extremely halting and deliberate. So he finally gave up and began communicating with only those who could understand him clearly through his thoughts.

I was much like that alien in trying to talk to that officer that night, and he certainly couldn't read my mind. Four years later a College Station officer pulled me over for driving 35 miles per hour in a 30-mph zone at 5 p.m.

"Little speedy today, huh?" the officer asked while strolling up to the car.

After a brief conversation, the officer said he'd write me a warning and to watch my speed. That warning marked a small victory for me of sorts. This officer never asked if I had a problem, and he never accused me of being drunk.

Ted with his brother, Pat, and his parents at Pat's wedding.

CHAPTER 15
My Brother's Wedding

I've never been embarrassed about having suffered a severe stroke and its resulting complications that still frequently leave me reaching for a word. Frustrated at times, yes, but never embarrassed. If I'd have allowed embarrassment to play a role in all of this I'd never have crept from the house.

Several such instances of frustration came at my brother Pat's wedding, which took place about six months after the stroke, in October 1993.

In the days and months following my stroke, Pat had really gone out of his way to help me out, by coming down from the Midwest immediately upon hearing of my ailment, and visiting and helping take care of me for an extended period.

Thirteen years separated us – I was the "influential" older brother – and when he was a kid I'd return home from college and take him to the movies or on trips during spring break.

When he was 9 years old, Pat bought a camera and projector with his earnings. His moviemaking continues today as an award-winning film and video director.

Pat offered extremely positive support in those days and months after the stroke, and I may not have recovered to this point without his help. Who knows how I might have responded if I'd have had a brother who was crying all of the time or who was sad because I was in this predicament.

A year before his wedding, in the fall of 1992, an anxious Pat had called me about his lovely girlfriend, Andi. Pat had a big-time question on his mind.

"Should I propose?" he asked simply and a bit nervously.

"If you feel strongly about her, I'd go for it," I responded, and Pat could tell I was smiling on the other end of the phone line.

Pat did propose (she said yes), and now here they were in a big Catholic wedding in the chapel of the St. Ursula Academy in Cincinnati. Only I wasn't in the best of shape to perform my duties as Pat's best man. I had flown from College Station to Dallas to Cincinnati, and was fairly worn out by the time of the rehearsal dinner on Friday night.

I wasn't the only miracle on two feet among the Odom boys that night at the rehearsal dinner, either. Pat had broken his neck three times in his life, twice in football and once in rugby.

"I'm very thankful," my father said at the dinner, "to have my two sons still alive."

During the rehearsal dinner at the University of Cincinnati Faculty Club, just about all of the guests stood up and told jokes or shared funny stories about the bride and groom.

As best man, I knew it was my job to tell the best story of all. Brooke had written out a humorous episode that had happened to Pat one time at the beach that I had dictated to her, and I prepared to share it with the others.

Finally, my turn came and I stood to tell the handwritten anecdote that certainly would be memorable among the guests, and also would show them I'd made quite a recovery from the stroke. But following a nearly 10-second pause out popped this bit of wisdom: "Ted Odom."

I quickly plopped down in my seat and stared at all of the faces staring back at me. Good thing they were friendly ones or I'd even been more uncomfortable at that point. Funny, but in that hugely frustrating moment I never once sheepishly looked down at the carpet.

There were some smiles and applause at the gesture, although some in the crowd probably didn't know I'd had a stroke. Friends and family figured this was a big step – to stand and say my name before a crowd – but from my end I was extremely disappointed. I had frozen and completely lost my train of thought, so how in the world would I ever stand before a college class and lecture for an hour?

Here I was, my brother's best man, and I couldn't even share one story? It was disheartening and a setback, or at least I thought so.

"That was the best speech," one relative said with a wide grin, "because it was the shortest."

I didn't get any better during the wedding ceremony, as twice I stood up at the wrong time, even though I had worked hard to memorize my cues. Later I realized everyone was focused on Andi and Pat, anyway, and not so much his bumbling brother.

During the wedding reception I quietly asked Pat's best friend, John Quinn, if he'd mind raising a toast to the bride and groom, and he gladly agreed.

My public performances that week left much to be desired, at least in my mind. Funny, though, everyone seemed happy with the effort, even if I figured after that rousing offering of, "Ted Odom," I might never speak to an audience again.

OTHER VOICES
Ed Odom, Father

Ted's toast didn't mean a nickel to me or the rest of the rehearsal dinner attendees, but it meant everything to Ted. He felt terrible because he believed it was a big test he didn't pass.

As a matter of fact his siblings and others gathered at the rehearsal dinner viewed Ted as miracle of sorts, based on his remarkable recovery over six months. We were proud of him for standing and saying his name.

Pat Odom, Brother

I chose Ted to be my best man as a testament to the friendship we always had. I also was glad that he was still alive after everything he'd gone through in the previous six months.

The best thing I felt I could do for Ted in that period is make sure he never felt all alone, and that he knew I was always thinking about him by picking up the phone and calling.

Those calls, of course, could be frustrating on both ends. Talking by phone eliminates all of the physical and visual cues that help the understanding of each other. Some might panic in that situation, if they believe they aren't communicating, but I learned to be patient and comfortable with the confusion. The bottom line was that even with a lack of understanding in the conversation between us, a phone call still helped Ted express himself to a loved one.

I also wanted him to know that he was still the same person to me, so I treated him the same as I always had. The problem for all of us, however, was realizing that Ted was never going to be the same person again. My family had to learn how to embrace the new Ted, while Ted at times tried desperately to cling to his past.

When I'd call he'd often recount experiences we'd shared before the stroke, and consecutive calls tended to drum up the same ol' stories. But after several years of this, my father spoke with Ted about how his holding onto the past was keeping him from his next steps in life.

That's when I saw several things change in Ted. His speech improved, he seemed more at ease and he began living for the moment more. He took pressure off of himself to be perfect, and he became less self-absorbed.

The earlier qualities may have been necessary to propel him through the speech therapy, but then the time came when he needed to remove the self-imposed stress and continue improving and healing.

Ted at a Christmas gathering with his mother, Toy, and brother, Pat.

CHAPTER 16
Willpower

Willpower is defined as the "control of one's impulses and actions."

Possessing willpower in my recovery came much easier through the strength and support of those around me, a support system I had virtually ignored before. What would have happened had negativity enveloped me in the days and months following the stroke? What if my family and friends had listened to the doctors who said I'd be "perfectly happy" in an institute?

You certainly wouldn't be reading this now. Never underestimate the power of positive words. I drew – and continue drawing – from the energy and encouragement of others.

My parents had faith that I'd recover enough in College Station that they didn't ship me off to St. Louis and unfamiliar surroundings. My siblings offered encouraging words through visits, cards and calls.

Brooke sat by my side as I drove for the first time six weeks after the stroke. Looking back, that took lots of faith. Would you ride with a man who had a severe stroke less than two months earlier?

Jean Foster never wavered during her lessons, and never let me believe anything other than I'd recover 100 percent.

My bosses at Texas A&M trusted that I would recover sufficiently enough to research and teach again, when they could have let me go based on what some doctors considered little chance of rehabilitation.

My church, A&M United Methodist, rallied to my side and provided the equivalent of a cane in easing me through the trying times.

The list goes on, and there's no room for negativity on it. As tough as it can be, avoid those who are only going to try and bring you down, because that's when you can stumble off the path of recovery.

Such positive feedback from those closest to me meant all the difference in my recovery, and supplied me with a tremendous amount of willpower to fight through the tough times.

Simply, those around me lifted me up in my most trying hour, and I'll always be thankful. Much of God's power is through the help of others – and helping others. When someone encourages you to follow through on a dream or a tough assignment or a further step in a recovery, that's God talking.

At that moment, you can do more than you ever knew. I'd have never made it without a support system, and someone even 100 percent healthy shouldn't isolate himself or herself from the love of those around them.

It's not easy to change negative behavior, for certain. There's not a switch you can flip and say, "All right, I'm gonna be holy now and have the willpower to fight temptation and do right in God's eyes."

Here's a few things that have worked and continue working for me:

* Set goals and then take action. Take small steps with such goals and once they are reached, then set more goals and further your development. Sometimes it's too overwhelming and we give up if trying to jump from A to Z while skipping B to Y.

In the 1991 movie, "What About Bob," starring Bill Murray, "Bob" takes "baby steps, baby steps" to reach small goals on the advice of his psychologist. It's a funny movie and good advice, even if Bob sometimes took the suggestion a little too literally.

* Read your Bible daily, even if it's for only 10 minutes.

* Attend church regularly – no man is an island and a supportive congregation provides a wonderful compass for a narrow path.

* Replace negative thoughts with their positive alternative. Being positive doesn't just mean saying positive things, it means focusing

on the good in everything. Sure, it's not realistic to believe that every thought that crosses your mind will be positive, but don't let negativity rule your day.

Understand that true strength comes only from God, and that Jesus taught us to pray this way: "Our Father, who art in Heaven, hallowed by the name. Thy kingdom come, thy will be done, on earth as it is in Heaven. Give us this day our daily bread, and forgive us our trespasses, as we forgive those who trespass against us. Lead us not into temptation, but deliver us from evil, for thine is the kingdom, and the power, and the glory, forever and ever, amen."

Please allow those holiest of words to sink in. The Lord's Prayer covers everything you need on this earth and paves the way for Eternal Life, and has been an invaluable staple in my recovery as a new and better man.

"God, you direct me, and I'll follow," I prayed to Him shortly after the stroke. "I'm at your mercy."

You must ask God to lead you on a new path of righteousness. Is such a walk easy? No. It's tough, but God's power will direct you, as His power has guided me. And with apologies to poet Robert Frost, that has made all of the difference.

OTHER VOICES
Ed Odom, Father

A tremendous amount of Jesus Christ's life was His teaching about faith. Christ had promised in Scripture that he would never leave us; that we wouldn't walk this world alone. This horrible thing perpetrated upon Ted was the ultimate example of the kind of faith we must have when seemingly overpowering adversity faces us.

Toy Odom, Mother

That first Christmas following his March 1993 stroke Ted told his father he'd wanted to read a short letter at the Epworth children's home that Ed runs.

"I want to thank everyone at Epworth's for your prayers for my recovery," Ted read from the letter at the Christmas dinner. "My recovery is not completely finished yet, and I will go forward with faith, God and a very good teacher. To the young people of Epworth, I want to say that I also have homework just about every day, and it's tough. But please work very hard to improve yourselves, just as I have to. Again, thank you."

I was really proud of Ted because he was able to focus in front of all of those people and read such nice words of encouragement. He was so glad to be alive, and he wanted to help others. We had never seen him do anything like that before. God had truly changed Ted's life through the stroke.

CHAPTER 17
Texas A&M: Home Away From Home

There are parts of Texas that are flat and dusty, but College Station, an hour northwest of Houston in the Southeast part of the state, is far from that stereotypical image of the Lone Star. Upon my arrival in 1981 I was positively surprised to encounter green grass and shady oak trees.

My Aggie colleagues' first impression of this brash Yankee weren't quite as pleasant back in '81.

"Hey, slow down, Ted!" fellow professor Fred Gardner yelled after me one fall morning in the Kleberg Center, home of the Poultry Science Department. "You walk way too fast. Why are you always in such a hurry?"

"I'll walk as fast as I want," I smugly said. "Try and see if you can keep up."

It seems truly a Northern thing to walk fast and go, go, go, because in Aggieland few folks seemed in a hurry to get anywhere. What was wrong with these people?

In graduate school I'd developed a brisk walk on the crowded paths at the University of Illinois, a method that also proved a solid psychological test. I strolled as fast as possible while staring down at the pavement, thus avoiding eye contact and certainly clearing a path.

People will practically jump out of your way if they believe you don't see them. "Cool," I thought with a sneaky grin, as nobody tried shouldering me off the sidewalk like the other students bothering with eye contact.

I never slowed until the stroke.

Upon my arrival in College Station I knew little about Aggie traditions, except that the place seemed like its own little world with such customs as Midnight Yell Practice before football games and Aggie Muster, a solemn ceremony honoring all Aggies who've died in the previous year.

One bright spring afternoon early in my tenure at A&M, I plopped down on a bench outside the Memorial Student Center, where a sign implores visitors to take their "Hats Off Please."

"The Memorial Student Center is dedicated to all Aggies who have given their lives in the defense of their country in any war, past or future," the sign reads. "In token of respect for their sacrifice, all individuals are requested to remove their hats in the building."

"Surely some rebel will break this rule in a heartbeat," I thought with a grin of the impending encounter I'd witness from the bench.

Sure enough, here came a big, surly looking cowboy, with faded Wrangler jeans and brown boots needing heavy polish. But as he grabbed the door to the MSC, off came the 10-gallon hat.

Next came a long-haired fellow in a baseball cap, and I practically rubbed my hands at the surely impending broken tradition.

"If anyone thinks the rules don't apply to him, it's a hippie," I gleefully thought while getting caught up in this new game of Test the Aggie Rituals.

Off came the Houston Astros cap.

"Whoa," I thought. "This place is for real."

Everything in Texas also was, "Y'all this," and "Y'all that." That took some getting used to after growing up on the phrase "you guys," but I adapted to the Southern address.

One thing I couldn't as easily adjust to: The way everyone touched you or patted you on the back. That part of Aggieland made me mightily uncomfortable. You just didn't do that where I came from, and I wasn't about to fit in.

"How are you, young man?" professor Bill Cawley typically boomed as his hand grasped my opposite shoulder – where my arm instinctively dropped like a little kid's trying to avoid his grandma's embrace – as Cawley unfailingly wrapped me in half a bear hug and furiously patted away.

"They're just different," I'd think of these frighteningly friendly Texans.

Any chance I got I hopped a plane back home to Champaign, because without adapting to this newfound friendliness, I was an outsider at Texas A&M. But that Aggie hospitality eventually played a huge role in my recovery. And following the stroke, of course, I finally felt comfortable putting my arm around a friend or squeezing the shoulder of another professor.

And thank goodness I met Linda, my first girlfriend in College Station, early on. Her connections and introductions to others kept me from leaving for good on several occasions. I never planned on staying for more than a few years, but now I've lived here a couple of decades.

In my first few years at A&M, my arrogance and insecurity had also turned off these otherwise mostly charming students to my delivery. That's when Dr. C.R. Creger, then the head of the Poultry Science Department, had called me into his office and told me I must change my teaching style to better relate to the students.

So it was change or get out, and luckily I made attempts to become a better and more affable professor, otherwise A&M as a whole might not have been so friendly and patient in the aftermath of my stroke.

I owe a debt of gratitude to Dr. Creger that words can't describe, because of the kindness he showed in the months and now years following the stroke.

He realized my homework with Jean Foster was of utmost importance, and total concentration on those simple lessons was invaluable in my recovery. Dr. Creger also set up my schedule to where I conducted 100 percent research, to remove the pressure of trying to return to the classroom.

The university could have shoved me aside a number of ways following the stroke that left me with a complete loss of language. But the good people of Texas A&M didn't – instead they embraced me when I needed it most – and for that I'll always be grateful.

That didn't mean there weren't hugely awkward moments along the way. Once my computer disk broke while I worked on an agricultural report that was part of a regional project, and I rushed down the hall to Dr. Creger's secretary, Jo Ann Pilkey.

"Jo Ann," I said as my panic slowly built. "My (disk) is broken." Only I didn't say "disk" – just a foul word that sounds much the same. Imagine the startled and then bemused stare in reply.

"JoAnn," I repeated for emphasis, wondering why she wasn't reaching in her desk for a new one, "My (disk) broke. Could I have another?"

Humor can be such a good thing in awkward situations, and that's about as clumsy as they come. Poor Jo Ann, once recovered from her initial shock, fought hard to keep from bursting into laughter while handing me a brand new computer disk. Had she giggled, I'd have giggled with her. All you can do is laugh in such situations. But JoAnn, like so many others at Texas A&M, treated me with much consideration in such a difficult time.

In the months before the stroke I had prepared letters applying for a full professorship at A&M. Several committees would decide if my 10 years-plus of service was worthy of the title – and a decent pay raise. Naturally the stroke postponed any such advancement. How could I have earned full professor status when I couldn't even say "full professor," or "Ted Odom"?

Three years after the stroke I once again prepared, along with the help of doctors Creger and Krueger, the papers for application for full professorship. Dr. Krueger was the head of the department when I took the job at A&M, and Dr. Creger became head of the department not long after.

In our preparation we didn't mention that I'd had a severe stroke a few years prior, we only discussed my teaching and research. Dr. Creger announced the promotion at the 1996 Christmas banquet, and the following September I began my duties as a full professor.

OTHER VOICES
Toy Odom, Mother

Ted's department head, Dr. Creger, knew he hadn't lost years of research tucked in his mind, and although he couldn't quite talk yet, Ted had worked so hard for the university in the years prior to the stroke. That work ethic clearly was one of the primary things that saved him later on. Otherwise how long would those in his department have felt compelled to wait for this man stricken by stroke?

Ed Odom, Father

The beautiful thing about Ted's story is how the university rallied to his cause, and how Dr. Creger stuck by him. A different kind of man, one more concerned about only the work in the department, would not have been as patient and might have found a way to get this problem out of his hair.

Such loyalty speaks highly of Ted's work before the stroke, because following Ted's setback we didn't know what kind of support the university would lend him. I didn't think Ted could get well quick enough to hang on to his job.

Texas A&M is amazing in that it is a massive family, and it's a beautiful thing to see. One time we visited A&M United Methodist Church with Ted, and the whole congregation yelled, "Whoop!" in the name of Jesus Christ.

"Whoop" is a celebratory Aggie yell.

If we hadn't lived in a university community in Champaign, we might have figured that's how folks in all college towns act. It's not. What other university in the world holds a ceremony like Aggie Muster, to remember every Aggie who has died in the past year?

Dr. Bill Krueger, Poultry Science Professor

When we were courting Ted after he graduated from the University of Illinois, he came down here for an interview and stayed several days, which we really liked. Being the head of the department, I tried hosting him the best way I knew how, even to the extent of taking him to the Dixie Chicken, a legendary watering hole in College Station. I bought him a greasy hamburger and a bottle of beer, and let him know the social life around here was pretty good.

I also knew the University of California-Davis was interested in Ted, and I expressed that to the dean and suggested that we make Ted a formal offer before he went out to UC-Davis for his interview.

We wanted him to have a solid job offer in his pocket when he ventured out West. That way, if he heard anything he didn't like, he already had a job waiting in College Station. He must have heard something he didn't like.

Ted arrived in College Station in the fall of 1981 and hit the ground running. His research skills proved invaluable. Following his stroke, Texas A&M stuck by him, and rightly so. He was sick and it was right to take care of him.

Honestly, his recovery has been remarkable. He could be a complete vegetable, but thanks to God, some good doctors and therapy, Ted leads a fairly normal life.

He's had to come back from scratch. While Ted lost much of the ability to remember many routine things – he would sometimes get lost in ordinary conversation – he was able to remember much of his more complicated research, which was an interesting phenomenon and a rarity. How you do that with the severity of the stroke he had is amazing, and the biology of all that is fascinating.

When Ted lost all of that ordinary stuff, of course, it changed his entire personality. When he came here as a faculty member, he was extremely impatient and impersonal – a guy always on the run.

And he was always hard on his graduate students. Ted didn't have a whole lot of time for other people, but since his stroke he's undergone a complete reversal. He shows a very high interest in others, to the point of regularly visiting people in the nursing homes and hospitals. And he's a much more religious person.

There's been quite a transformation during Ted's recovery. In my case, when I had my first set of heart bypass operations in January 1973, Dr. Denton Cooley – the famed heart surgeon who performed my operations – strolled into my room one day during my recovery at St. Luke's Hospital in Houston. I was 51 years old.

"Mr. Krueger," he said, "I really don't know how you lived to be 50, because 99 percent of the people who had the kind of heart blockage that you had are dead. The only thing I can figure is the Good Lord has got something planned for you to do before you die."

I think about that statement every day, and Dr. Cooley passed that wisdom along three decades ago. It changed my attitude toward life. I imagine, too, that God had something special planned for Ted Odom back in 1993, as He does today.

CHAPTER 18
Scotland: Mission Possible

Nearly 16 months following the stroke I flew to Glasgow, Scotland, in August 1994 to attend the European World Poultry Science Conference.

I'd been invited to the prestigious meeting before the stroke, and I figured the trip provided an outstanding gauge of my recovery.

Plus, there were many people from around the world who work on different parts of the heart who'd attend the meeting, and this was a wonderful opportunity to meet them and share information. I wasn't about to cancel, despite concerns this might be too much too fast in my recovery. I wanted to see if I could do this, for myself and for the department.

The meeting was to take place at a large conference center in Glasgow, just west of the Roslin Institute near Edinburgh, the same place where two years later in July 1996 scientists would clone a sheep, the infamous "Dolly."

I'd worked extremely hard on my presentation about heart failure in chickens, and to prepare I made slides so that I could read the text aloud as I jumped from one image to the next.

While finding the right words often still troubled me, I figured I could memorize the text and the accompanying pictures would stimulate my brain in remembering the words. And if I couldn't find a word, I was determined to just "slide" on to another part of the presentation, and no one would know I'd skipped parts of the text.

In the months leading up to the trip, I even took a slide projector home so I could practice there, all alone and into the darkness that enveloped the College Station nights.

During the day I'd practice in the Kleberg Center, home of the Poultry and Animal Science Department. It featured a huge auditorium I figured might resemble the one in Scotland.

I'd practice talking to the left side and the right side and very close to the projector. While such preparation was never a thought prior to the stroke, now it was extremely important that I cover all of the angles in returning to the front of the class.

On a Friday afternoon, I flew from College Station to Dallas, and then on to Chicago. In the bathroom before the flight to Glasgow, I ran into movie critic Roger Ebert. He was Class of 1960 at Urbana High, and I had graduated nine years later. He was just in from Scotland, returning from a big film festival in Edinburgh.

After informing him I also hailed from Urbana, Ebert laughed and said he missed the Steak & Shakes native to our fair town.

It was a long, overnight flight to Scotland and the following evening I tried resting as much as possible before the opening of the meeting on that Sunday night, but of course this was difficult because of the anxiety attached to such a big event in my life.

That Sunday afternoon colleagues began gathering in the convention center, and some seemed shocked to see me.

"I thought you had a serious stroke," one friend said, his eyes in disbelief.

"Yes, I did," I replied, and following a few more words he realized I had some sort of impairment because of my slow, deliberate speech.

"Well, I guess the stroke wasn't that serious," one colleague said while making small talk.

"Yes, it was," I said, nodding my head emphatically. "But I'm here to present my data."

I'd say losing your language is pretty serious.

No one in Scotland knew, of course, if I'd ever recover enough to teach again or research, and the doubt oozing from several of my colleagues practically permeated my being during those few days overseas. But while I had lost the language, I still had my research ideas and theories.

I couldn't talk, write or read in the weeks following the stroke, but the previous knowledge I'd absorbed still lay intact, and I'd already began drilling into it like a recently rediscovered oil field.

In the midst of the many questions – and sometimes sense of skepticism about the stroke's severity – I grew more and more nervous about my impending presentation.

I hadn't tried hiding the fact that I suffered from aphasia – the inability to remember words – but one of the hosts and the man who invited me, Dr. Martin Maxwell, hadn't realized how much I had lost. On the morning of my presentation I finally chatted briefly with Dr. Maxwell and he soon realized the severity of my impairment.

In the hour leading to my presentation, I paced the floor nervously liked a freshly-caged wildcat, trying to quash any bewilderment at the predicament I'd placed myself in on the road to recovery.

I presented first among all of my colleagues, and as I spoke the faces of the 100 or so in attendance seemed ashen, contorted and disfigured. I fought off the feelings that I was suddenly in the third grade again, standing before a class of critical mates having left my book report at home – and that I'd forgotten to put on my pants to boot.

"Mr. Chairman, members of the World Poultry Science Association, and guests, I want to begin by saying that I have a hypertrophic heart problem," I said. "My heart pumped out blood clots and pushed them into my brain. That's why I had a stroke 16 months ago. The next slide you'll see is a CAT scan image of my brain. I didn't have lesions in the motor areas so I wasn't paralyzed."

During the course of the 30-minute presentation, all of that pacing and fretting had used a ton of my energy. I practically ruffled the hair-dos of my colleagues with the sigh I exhaled once I was done.

"Thank you, Dr. Odom, for that nice presentation," Dr. Maxwell said.

"I can answer any questions," I said, sweat beading on my furrowed brows.

I then fielded several questions – ones anyone in my position might normally field – but I couldn't understand a couple of them because even though they were asked in English, I had trouble with some of the international accents.

So I simply wouldn't say anything when faced with a question I didn't understand, and Dr. Maxwell answered for me. I was so fatigued at that point I seemed completely impaired again. It reminded me of the time I missed my cues at Pat's wedding, and that same helpless feeling swept over me like high tide at the beach.

I presented on a Monday morning, and the meeting ran through Thursday. I tried listening to others, but hearing such complex theories overwhelmed me, and at one point I figured I might be in for another stroke. So I just kept my distance for a time.

On Wednesday night, Dr. Maxwell invited me to dinner at his home. While passing the hors d'oeuvres, Dr. Maxwell said, "Ted, I wasn't aware you'd suffered so much damage."

I appreciated his sincerity and we discussed my situation for more than an hour before shaking hands and agreeing to see each other at the next meeting.

Many of my colleagues at the meeting were complimentary that I'd worked up the courage to attend and present at the meeting, still others acted as if I had died by simply ignoring me.

As I climbed on the plane bound for the United States, I knew I'd have to continue working hard on learning to speak in front of groups again, because this no longer was an automatic thing. But I also learned from the experience that faith can carry you through some tough experiences, and presenting in Scotland before my peers marked a big step in my recovery.

Among all of the questions at the conference, the perplexed looks, some heads turning the other way and my own anxieties, at that point only God knew I'd be back in the classroom for sure.

CHAPTER 19
A Leap Back to Lectures

OTHER VOICES
Ed Odom, Father

The thing that seemed to drive Ted the most in the aftermath of the stroke was the desire to be a professor again. Teaching and what it means to the students were really important to him and he wanted that back.

What he observed from older people is they get as well as they want to get. Getting well is what Ted desired most, and he had an overpowering hunger to become what he was.

Perhaps Ted could have recovered his ability to research without being a communicator, but the idea of returning to the head of the classroom, if only for a time, supplied him with an impetus to get better.

TED ODOM

There's no perfect script to follow for a person recovering from a stroke. There's no overnight recovery, and improvement from brain damage usually won't happen in the span of a two-hour movie, as it did in Harrison Ford's excellent "Regarding Henry."

But the story here is I returned to the classroom to teach again, despite a severe setback in the prime of my life. My task was rebuilding my language so I could work in the department again, and show it could be done.

And on a glorious spring day in 1995, I'd made it back to the head of the classroom, and was focusing hard on a variety of diagrams on the chalkboard and a lecture on a chicken's cardiovascular system when a student accidentally – I presume – passed gas.

Even that bit of uneasiness hardly deterred me from the task at hand, until the muffled giggles continued.

"All right," I said, turning from the chalkboard and breaking from an intense concentration. "Whodunnit?"

Believe it or not, I've got that student to thank for helping ease my initial tension. It was my first return to teaching since the March 1993 stroke, and as I stood before about 30 undergraduate students enrolled in Poultry Science 202 I calmly reminded myself that this is what I had worked for.

This day, April 21, 1995, also marked the second anniversary of my first class with Jean Foster.

"I hope I can help you learn," I told the attentive students. "I'm going to talk about the heart and how it pumps. I had a stroke more than two years ago, so this is personal for me. I had a blood clot in my brain, and the lack of oxygen because of that clot killed that part of the brain.

"If I lose my thoughts or pause while searching for a word, it's a result of this disability. But I'm still going to try and help you learn."

I was a guest lecturer in the class, and following an hour's worth of speaking on the chicken heart and its functions – and how it compares to the human heart – I received extremely positive student feedback.

Even had a student absorbed every word that day, there's no way he or she could have learned as much as I did from that lecture. It was true. All things are possible through God. Two years prior I hadn't even known my name because of a debilitating stroke. And now it gave me great joy to pass along knowledge to kids eager to learn.

***

The following fall a relapse of sorts kept me out of the classroom. That September I was in the College Station mall and suffered what doctors later said was a panic attack.

I hurried to my car and tried to drive but couldn't, and so I sat dazed in the parking lot for more than an hour. Once I got my bearings together enough to drive, I finally arrived home and fell into a fitful sleep. I spent the next day in the hospital and the symptoms – double vision, nausea, tunnel vision and dizziness – finally faded.

That scary episode removed me from the classroom that fall and then the spring of 1996 – similar instances would occur along the way but none so severe – and the following fall I lectured a couple of times to an upper-level Poultry Science class, Avian Physiology.

That week I also helped the nearly 40 students in the class in their afternoon labs, and it reminded me of my undergraduate years at Eastern Illinois, when I helped fellow students in Cat Anatomy. For the first time in years, I enjoyed that same rush of adrenaline from teaching others that I'd first felt back at Eastern Illinois.

Because of the exhausting nature of the morning lectures, it was difficult for me to talk during the afternoon labs, but the students never laughed or made fun of me. They understood and were sensitive to the idea that I was working hard to teach them about the hearts of birds, and for that I'm still grateful.

During one of the morning lectures, I was explaining the nature of the damage to my brain with the aid of a transparency on an overhead – or so I thought. Several students cleared their throat to let me know that I'd placed a piece of paper on the overhead, not a transparency, and that the screen I supposedly was teaching on was pitch black.

When I looked up at the screen, I did my best impression of "Kramer" from the TV show "Seinfeld" by kicking up my legs and nearly falling backward. I had made light of my forgetfulness and the students responded with good-natured laughter.

I conducted 100 percent research through the spring of 1997, and that fall I once again gave a couple of lectures and helped with the labs of Avian Physiology. This time was much better, in fact, following another year of recovery. My speech was much more fluid and I had many less slip-ups.

So finally, in the spring of 1998, I taught 10 lectures to a graduate course, Poultry Science 609, an advanced Avian Physiology course. Now, these were two-hour classes and by lecture's end I was usually drained, as were the students. I'd usually take a short break 40 minutes into the lecture and then return to finish up.

I also worked hard to be a better teacher than I had before. In the years following my stroke I had begun regular Sunday visits to the hospital and nursing homes, to call on friends who needed someone to talk to – and listen.

Now, those visits were paying off because they helped distinguish the important things in life. I no longer felt a tremendous urge to be a perfectionist again in my lectures. I only wanted to teach the students the best way I knew how.

Part of that meant trying to think from the students' point-of-view. After all, I'd been a student myself in the couple of years prior under the tutelage of Dr. Jean Foster. If sometimes students didn't act like they wanted to understand or listen, I practiced patience. When you've recently been behind the desk instead of in front of it, you get a better idea of how the students feel.

I also valued good writing more than ever before. For years prior to my stroke I'd sail through grammatical errors, and fire off grants and proposals without even asking a peer to review them.

One such proposal returned with a note attached: "Good ideas, but this guy can't write."

The critic was right, even if I denied it at the time. I've learned the clearly written word is an invaluable communication tool.

I'd call returning to the classroom to teach a big high, but that's not the right word. A better term may be "happy." Because with a high you're always searching for another high and when you don't get it, you're depressed.

I'd fulfilled a huge personal goal of teaching again, along with conducting plenty of research. In November 2005, I retired from the university, and now spend much of my time visiting folks in nursing homes.

Influenced by Pastor Malcolm Bane of College Station Baptist Church, who had visited me in the hospital two days after my stroke, I had starting visiting nursing home residents in 1996. It means so much to people to have company – someone simply to talk to – in their times of need.

I also still quench my desire to teach by working with other stroke victims on their path to recovery. And I pray that by day's end they know a little bit more than before we began.

CHAPTER 20
God's Gift: A Priority Change

OTHER VOICES
Father, Ed Odom

Ted's pre-stroke world was awfully narrow. He was much more interested in his own life than anyone else's. The stroke brought about a drastic change in his personality, as he evolved from being bent on conquering the world to slowing down and noticing things like wildflowers along the highway and their inherent, simple beauty.

Ted had never bothered catching a sunrise or sunset before, and now those glorious wonders are a part of his beautiful new world. Ted also makes many more simple but profound gestures like phone calls to his siblings and other relatives, usually only to ask a sincere "How are you?" None of this would have happened without his having had a stroke at age 41.

TED ODOM

Enduring a life-altering stroke was the best thing that ever happened to me.

The experience itself, of course, was brutal beyond comprehension, one comparative to someone losing a physical extremity. The phenomenon of phantom limbs occurs when nerve endings are still there, and that person has got an echo in the motor area of his brain, which believes the limb still exists. This is a false stimulation, and one

103

that I compare to the phantom, weird words that following the stroke bounced around in my brain, words with little meaning to a man who'd lost his language.

Despite such a once-terrifying state, I've learned to use what happened in a positive way, and I've learned to share God's grace through that life-changing experience. Simply, my old lifestyle hadn't cut it, and offered only emptiness in the midst of those many lonely nights of self-examination that had usually followed what was supposed to be a "good time."

A somewhat sad part of this entire experience meant losing some friends that I had previously spent a lot of time with, as we used a better part of our hours together drinking and telling crude stories. But following the stroke I no longer desired those things and our personalities didn't mesh.

Ted hangs out with brother Pat and nephew Peter.

Signs of God's love continue reminding me I've made the right decision in trying to live my life through the Bible. Bluebonnets, for example, are a Texas staple – and the state's official wildflower – but they don't typically grow in suburban backyards. On my first trip to Houston for medical treatment following the stroke, I clearly remember waves of bluebonnets enveloping Highway 6 on the way down from College Station.

When I returned from Houston and began my physical, emotional and spiritual recovery, that next season an abundance of bluebonnets sprouted forth for the first time in my small backyard. I believe they signaled a wink from God letting me know everything was going to be OK – in fact, better than OK.

I never thought much about bluebonnets before, but the dazzling details of that little flower are amazing – and God is in the details.

I used to get upset with anything less than perfection, but now I'd rather not. Remember Coach Warren "Smitty" Smith from my old football days at Urbana High? He used to hound his players with promises of, "If you mess up I'll follow you forever. I'll be in your head forever."

He'd shout and yell so much that I remember it still – and in a way, I guess he was right. He's still occasionally in my head, his stern vows ringing clear, but now I know it's OK to mess up occasionally.

I never wanted to start down that trail in teaching once I was back in the classroom – one where threats and intimidation formed the basis for learning. The man waging that war needs to reevaluate his priorities, and I was that man before my comeuppance in New Orleans. You must practice constantly to change your behavior – it certainly doesn't come automatically – and pitfalls are always there for a slip back to the old way of doing things.

But I've also learned to relax, try and do right and let God's will be done. I don't always know exactly how to handle certain situations or what I should do, but God always knows. A big part of my recovery came in realizing that only God is directing my life, and that negative energy – which through Satan will try and soak your every fiber – must be discarded with a simple prayer of, "Thy will be done."

And when I pray and meditate, God's answers ring loud and clear. The more I read and reread the Bible, the better I understand God's plan for us – and how it relates to my life. As Jesus taught us: If birds don't worry, then why should we? God feeds the birds, and God will feed us.

You may not believe you have time for prayer and meditation because of an exceedingly busy schedule, and that's exactly what Satan wants you to think. There are forces working against you each day that will try and keep you from daily prayer and reading the Bible, but please take the time for God. He's taken the time for you.

***

I revisited New Orleans four years after my stroke, and even rented a hotel room at the de la Poste, for old times' sake and for a bit of closure. My friends and I also visited Bourbon Street again but I said, "No, thanks," to the alcohol, as I've done since the stroke. But, boy, I sure loved listening to those lively Zydeco tunes echoing through the streets. God, it's certainly true, is in the details of wildflowers and even upbeat music.

Mine positively isn't a perfect story – only Jesus' life set a perfect example – and I'm not an amazing person. God is amazing. Before the stroke He was trying to reach me through my busy lifestyle to let me know that He needed me for His work, and that I needed to focus less on my material surroundings.

Certainly, I shallowly communicated with people all of the time beforehand, although even I didn't understand most of my own reasoning for such a secular, barren life. But when I became a prisoner within myself, with no language to share with others, I veered onto life's right, narrow path and now hope to share God's Word with every ounce of energy He's providing me.

God sent me back that bright March day in New Orleans, with a crystal-clear purpose for the rest of my time on earth and a contract – The Holy Bible – to spread His Word. Certainly, I could say, "Thanks, God, for sparing me," and just continue the life I had led before.

But I let Jesus direct my path daily, and He's wholeheartedly welcomed me into His loving, eternal embrace. Besides, reverting to the old way of doing things isn't in the contract.

In the end, this devastating stroke made me a better teacher. And, thankfully and most importantly, a better student of God's amazing grace.

About Brent Zwerneman:

Brent Zwerneman is a senior sports writer for the *San Antonio Express-News*, where he covers Texas A&M University and the NFL's Houston Texans. Zwerneman has previously written "Game of My Life: 25 Stories of Aggies Football" and "More Tales From Aggieland" for Sports Publishing LLC.

He covered the Aggies for four years (from 1995-99) as a columnist for the *Bryan-College Station Eagle*, winning three Katie Awards from the Press Club of Dallas and two Headliner's Awards from the Press Club of Austin in that span. Zwerneman has since covered the Dallas Cowboys and helped cover the San Antonio Spurs for the *Express-News*.

Zwerneman was a first-team all-district basketball player at Class 4A Oak Ridge High School, north of Houston, and he then played college baseball at Sam Houston State University, where he was a pitcher. He also served as editor of the university's *Houstonian* newspaper, which won first place in the state under his guidance.

Zwerneman later took a couple of postgraduate classes at A&M, before shelving the ivory tower teachings for book writing in his spare time. Besides covering a couple of Super Bowls and the Olympics, two of Zwerneman's claims to Aggie fame are playing a little catch in the off-season with John Scheschuk, the 1999 A&M baseball squad's team captain, and being cousins with former Aggies standout track athlete Rob Graf.

Zwerneman can be reached at brentzwerneman@hotmail.com.

Printed in the United States
56020LVS00009B/43-48